FUNDAMENTALS OF CAPACITANCE & INDUCTIVE AND CAPACITIVE REACTANCE

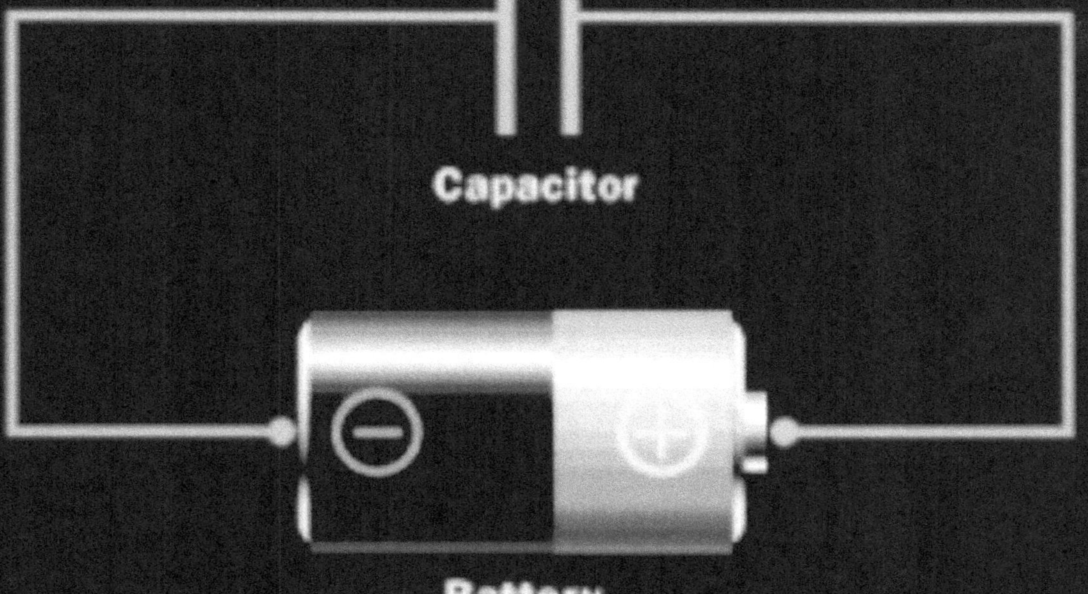

Capacitance

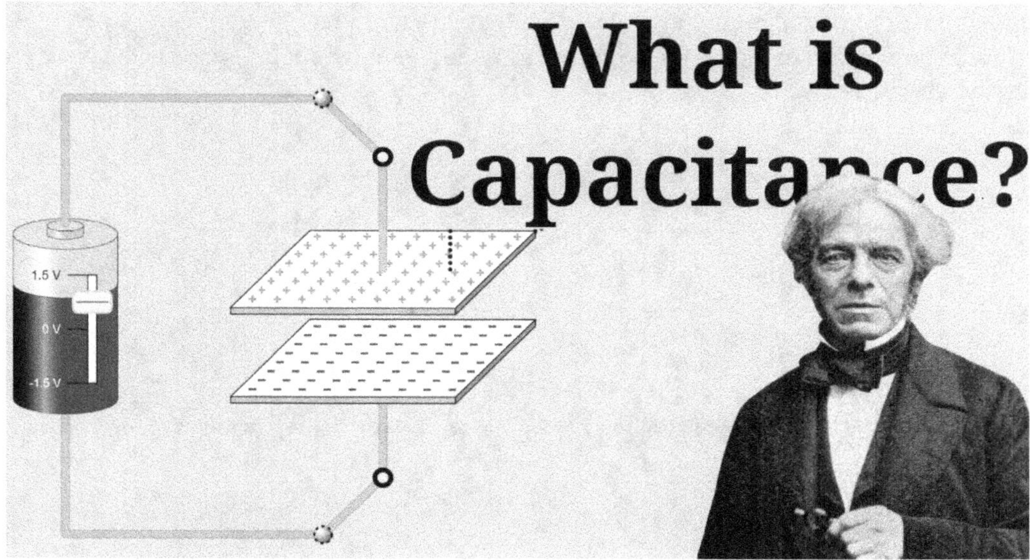

Capacitance is the ability of a device to store electric charge. It is a measure of the amount of electric charge that a device can hold per unit of voltage applied to it. In other words, it is a measure of the device's ability to store energy in the form of electric fields.

Capacitance is defined as the ratio of the charge stored in a device to the voltage applied to it. Mathematically, it can be represented as:

Capacitance (C) = Charge (Q) / Voltage (V)

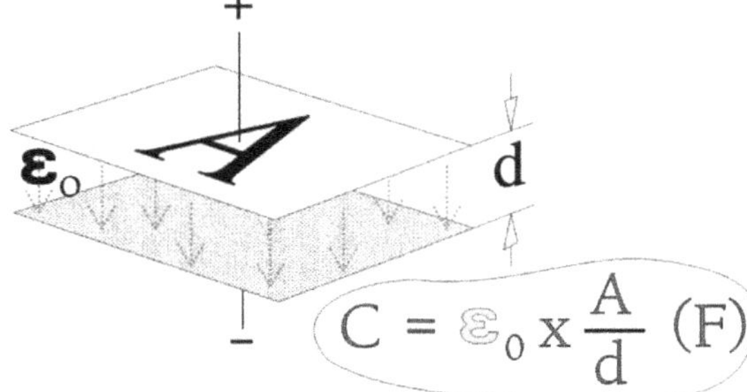

The unit of capacitance is the Farad (F), which is defined as the charge in Coulombs that a capacitor can store per volt of potential difference between its plates.

In practical applications, capacitors are used to store electric charge and release it as needed. They are commonly used in electronic circuits to filter or regulate voltage, store energy for short-term power backup, and block DC voltage while allowing AC signals to pass through.

The formula for capacitance can be derived from the equation for electric field, which is:

$E = Q / \varepsilon o A$

where E is the electric field strength, Q is the charge, εo is the permittivity of free space, and A is the area of the capacitor plates.

By rearranging this equation, we can get:

$Q = \varepsilon o A E$

Now, if we divide both sides of this equation by the voltage (V) applied across the capacitor, we get:

$C = \varepsilon o A / V$

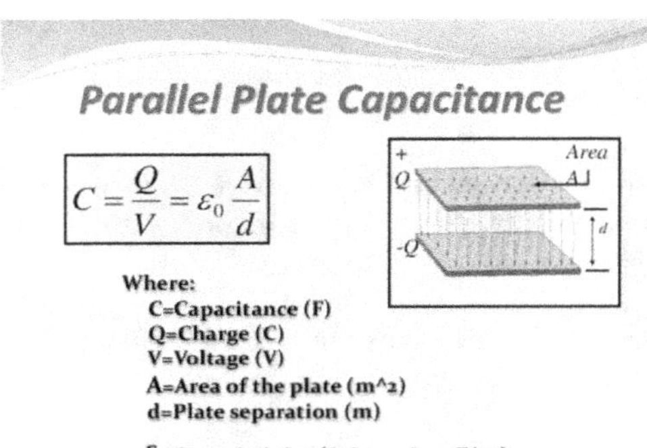

This is the formula for capacitance, where εo is the permittivity of free space (8.85 x 10^-12 F/m), A is the area of the capacitor plates, and V is the voltage applied across the capacitor.

Capacitance is a measure of the amount of electric charge that a device can hold per unit of voltage applied to it. It is defined as the ratio of charge to voltage, and its unit is the Farad (F).

Yes, you are correct. Capacitance is indeed measured in Farads (F), and it represents the ability of a capacitor to store electric charge. The capacitance value indicates the amount of charge that can be stored per unit voltage.

The basic formula for capacitance is:

$C = Q / V$

Where C is the capacitance in Farads (F), Q is the charge in Coulombs (C), and V is the voltage in Volts (V).

In addition to the basic formula, there are other formulas and concepts related to capacitance:

1. Capacitance of a parallel plate capacitor:

C = εo * A / d

Where C is the capacitance, εo is the permittivity of free space (approximately 8.85 x 10^-12 F/m), A is the area of the plates in square meters (m^2), and d is the separation distance between the plates in meters (m).

2. Capacitance of capacitors in series:

The total capacitance of capacitors connected in series is the reciprocal of the sum of the reciprocals of individual capacitances:

1 / C = 1 / C1 + 1 / C2 + ... + 1 / Cn

Where C is the total capacitance, and C1, C2, ..., Cn are the individual capacitances.

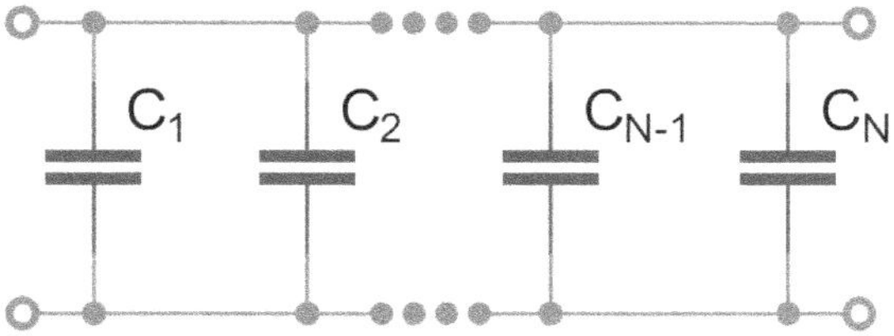

$$C_{Tot} = C_1 + C_2 + \ldots + C_{N-1} + C_N$$

3. Capacitance of a capacitor with a dielectric material:

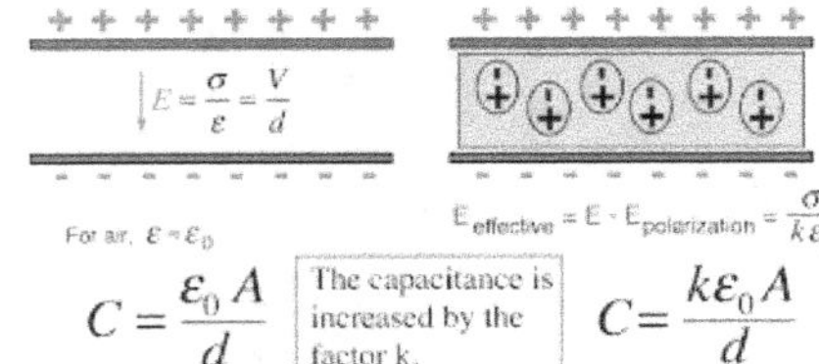

When a dielectric material is inserted between the plates of a capacitor, it increases the capacitance. The formula for the capacitance with a dielectric material is:

C = εo * εr * A / d

Where C is the capacitance, εr is the relative permittivity (also known as the dielectric constant) of the material, and all other variables have the same meanings as before.

4. Capacitance of a spherical capacitor:

For a spherical capacitor, the capacitance is given by:

C = 4πεo * A / d

Where C is the capacitance, εo is the permittivity of free space, A is the surface area of the inner or outer sphere, and d is the distance between the two spheres.

These formulas and concepts provide a framework for understanding and calculating capacitance in various configurations.

Here are some additional details and concepts related to capacitance:

1. Energy Stored in a Capacitor:

A capacitor can store energy in the form of electric potential energy. The energy stored in a capacitor can be calculated using the formula:

E = (1/2) * C * V^2

Where E is the energy stored in Joules (J), C is the capacitance in Farads (F), and V is the voltage across the capacitor in Volts (V). This formula shows that the energy stored in a capacitor is proportional to the square of the voltage across it.

2. Time Constant:

The time constant of a capacitor-resistor (RC) circuit is a measure of how quickly the capacitor charges or discharges. It is defined as the product of the resistance (R) and the capacitance (C):

τ = R * C

The time constant represents the time it takes for the voltage across the capacitor to reach approximately 63.2% of its final value during charging or discharging.

3. Dielectric Materials:

Dielectric materials are insulating materials that can be inserted between the plates of a capacitor to increase its capacitance. When a dielectric material is present, the formula for capacitance becomes:

$$C = \varepsilon_o * \varepsilon_r * A / d$$

Here, εr (relative permittivity or dielectric constant) is a dimensionless quantity that represents the ability of the material to store electric charge. The presence of a dielectric material increases the capacitance by a factor of εr.

4. Types of Capacitors:

There are various types of capacitors used in electronic circuits, each with its own characteristics and applications. Some common types include:

- Ceramic capacitors: These are compact and inexpensive capacitors with a wide range of capacitance values. They are commonly used in electronic circuits.

- Electrolytic capacitors: These capacitors have a higher capacitance value and are polarized, meaning they have a positive and negative terminal. They are commonly used in power supply circuits.

- Film capacitors: These capacitors are made from a thin dielectric film and are available in a variety of capacitance values. They are used in various applications, including audio circuits.

- Tantalum capacitors: These capacitors have a high capacitance value and are polarized. They are commonly used in applications requiring a compact size and high capacitance.

These are just a few additional details and concepts related to capacitance. Capacitors are fundamental components in electronic circuits and are utilized in a wide range of applications.

****

Capacitive and Inductive Reactance

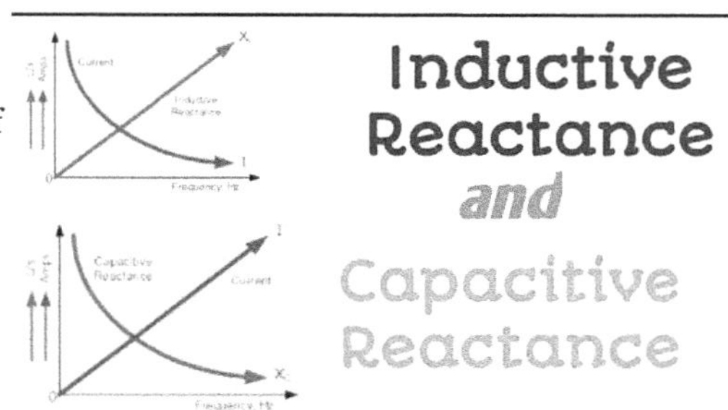

Capacitive reactance and inductive reactance are two types of impedance that arise in AC (alternating current) circuits due to the presence of capacitors and inductors, respectively. Both capacitive and inductive reactance are frequency-dependent and affect the flow of current in the circuit.

1. Capacitive Reactance (Xc):

Capacitive reactance arises in circuits that contain capacitors. It is denoted by Xc and is defined as the opposition to the flow of AC current through a capacitor. The formula for capacitive reactance is:

Capacitive Reactance

AC

C

$$X_C = \frac{1}{\omega C} = \frac{1}{2\pi f C}$$

Xc = 1 / (2πfC)

Where Xc is the capacitive reactance in ohms (Ω), f is the frequency of the AC signal in Hertz (Hz), and C is the capacitance of the capacitor in Farads (F).

Capacitive reactance decreases as the frequency of the AC signal increases. At low frequencies, capacitive reactance is significant, and it reduces the flow of current. As the frequency increases, capacitive reactance decreases, allowing more current to flow through the capacitor.

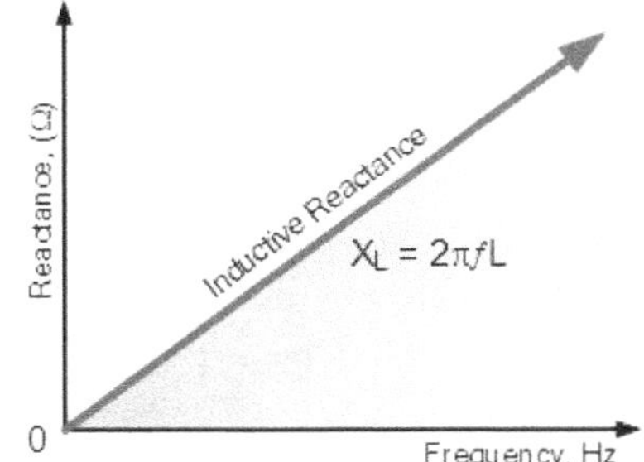

2. Inductive Reactance (Xl):

Inductive reactance arises in circuits that contain inductors. It is denoted by Xl and is defined as the opposition to the flow of AC current through an inductor. The formula for inductive reactance is:

Xl = 2πfL

Where Xl is the inductive reactance in ohms (Ω), f is the frequency of the AC signal in Hertz (Hz), and L is the inductance of the inductor in Henries (H).

Inductive reactance increases as the frequency of the AC signal increases. At low frequencies, inductive reactance is relatively small, and it allows more current to flow through the inductor. As the frequency increases, inductive reactance increases, limiting the flow of current through the inductor.

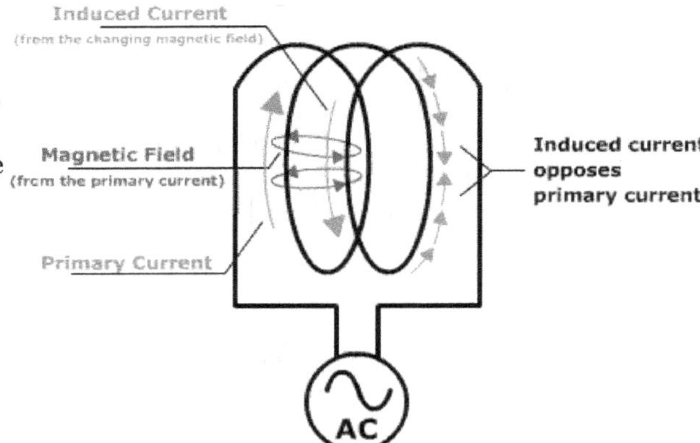

Inductive reactance is proportional to both the frequency and the inductance of the inductor. It plays a crucial role in circuits with inductive components, such as transformers, motors, and solenoids.

It's important to note that both capacitive and inductive reactance are part of the overall impedance (Z) of an AC circuit. Impedance is the total opposition to the flow of AC current and is calculated as the vector sum of the resistive component (R) and the reactive components (capacitive reactance and inductive reactance):

$$Z = R + j(Xl - Xc)$$

Where j is the imaginary unit (√(-1)). The combination of resistive and reactive components determines the behavior of the AC circuit and affects the phase relationship between voltage and current.

Aadditional details and concepts related to capacitive and inductive reactance:

1. Phasor Representation:

Capacitive and inductive reactance are represented as complex numbers in AC circuits. These complex numbers are often expressed using phasors, which are vectors that represent the magnitude and phase of the impedance.

In the phasor representation, capacitive reactance is represented by a negative imaginary number (-jXc), while inductive reactance is represented by a positive imaginary number (jXl). The sign convention is based on the phase relationship between voltage and current in capacitors and inductors.

The impedance (Z) can be expressed as a complex number using phasors:

$Z = R + j(Xl - Xc)$

The magnitude of the impedance is given by $|Z| = \sqrt{(R^2 + (Xl - Xc)^2)}$, and the phase angle ($\theta$) is given by $\theta = \text{atan}((Xl - Xc) / R)$.

2. Capacitive and Inductive Reactance in AC Circuits:

In AC circuits, the presence of capacitors and inductors can affect the behavior of the circuit. Capacitive reactance and inductive reactance interact with the applied voltage to create phase shifts between voltage and current.

- Capacitive Reactance: Capacitors block DC (direct current) but allow AC (alternating current) to pass through. Capacitive reactance decreases as the frequency increases. At very high frequencies, the capacitive reactance approaches zero, allowing AC current to flow almost unimpeded. Capacitors cause a leading phase shift between voltage and current, with current leading voltage by 90 degrees.

- Inductive Reactance: Inductors allow DC to pass through but impede the flow of AC. Inductive reactance increases as the frequency increases. At very high frequencies, the inductive reactance becomes significant, limiting the flow of AC current. Inductors cause a lagging phase shift between voltage and current, with current lagging voltage by 90 degrees.

The presence of capacitive and inductive reactance in an AC circuit can cause the voltage and current to be out of phase with each other, leading to complex behaviors such as resonance, power factor correction, and filtering.

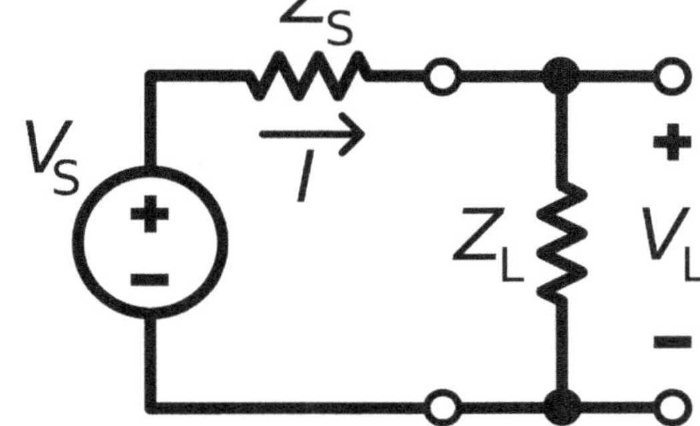

3. Impedance Matching:

Capacitive and inductive reactance play a role in impedance matching, which is the process of optimizing the transfer of power between different components or systems. In some cases, capacitors and inductors are used to adjust the impedance of a circuit to match the impedance of the connected components, ensuring maximum power transfer and minimizing reflections.

For example, in RF (radio frequency) circuits, capacitors and inductors are often used in filter networks or matching networks to match the impedance of antennas, amplifiers, or transmission lines.

These are some additional details and concepts related to capacitive and inductive reactance. Understanding the behavior of capacitors and inductors in AC circuits is essential for designing and analyzing circuits involving AC signals.

CAPACITANCE

LEARNING OBJECTIVES

Upon completion of this chapter you will be able to:

1. Define the terms "capacitor" and "capacitance."
2. State four characteristics of electrostatic lines of force.
3. State the effect that an electrostatic field has on a charged particle.
4. State the basic parts of a capacitor.
5. Define the term "farad".
6. State the mathematical relationship between a farad, a microfarad, and a picofarad.
7. State three factors that affect the value of capacitance.
8. Given the dielectric constant and the area of and the distance between the plates of a capacitor, solve for capacitance.
9. State two types of power losses associated with capacitors.
10. Define the term "working voltage" of a capacitor, and compute the working voltage of a capacitor.
11. State what happens to the electrons in a capacitor when the capacitor is charging and when it is discharging.
12. State the relationship between voltage and time in an RC circuit when the circuit is charging and discharging.
13. State the relationship between the voltage drop across a resistor and the source voltage in an RC circuit.
14. Given the component values of an RC circuit, compute the RC time constant.
15. Use the universal time constant chart to determine the value of an unknown capacitor in an RC circuit.
16. Calculate the value of total capacitance in a circuit containing capacitors of known value in series.
17. Calculate the value of total capacitance in a circuit containing capacitors of known value in parallel.
18. State the difference between different types of capacitors.
19. Determine the electrical values of capacitors using the color code.

CAPACITANCE

In the previous chapter you learned that inductance is the property of a coil that causes electrical energy to be stored in a magnetic field about the coil. The energy is stored in such a way as to oppose any change in current. CAPACITANCE is similar to inductance because it also causes a storage of energy. A CAPACITOR is a device that stores electrical energy in an ELECTROSTATIC FIELD. The energy is stored in such a way as to oppose any change in voltage. Just how capacitance opposes a change in voltage is explained later in this chapter. However, it is first necessary to explain the principles of an electrostatic field as it is applied to capacitance.

Q1. Define the terms "capacitor" and "capacitance."

THE ELECTROSTATIC FIELD

You previously learned that opposite electrical charges attract each other while like electrical charges repel each other. The reason for this is the existence of an electrostatic field. Any charged particle is surrounded by invisible lines of force, called electrostatic lines of force. These lines of force have some interesting characteristics:

- They are polarized from positive to negative.
- They radiate from a charged particle in straight lines and do not form closed loops.
- They have the ability to pass through any known material.
- They have the ability to distort the orbits of tightly bound electrons.

Examine figure 3-1. This figure represents two unlike charges surrounded by their electrostatic field. Because an electrostatic field is polarized positive to negative, arrows are shown radiating away from the positive charge and toward the negative charge. Stated another way, the field from the positive charge is pushing, while the field from the negative charge is pulling. The effect of the field is to push and pull the unlike charges together.

Figure 3-1.—Electrostatic field attracts two unlike charged particles.

In figure 3-2, two like charges are shown with their surrounding electrostatic field. The effect of the electrostatic field is to push the charges apart.

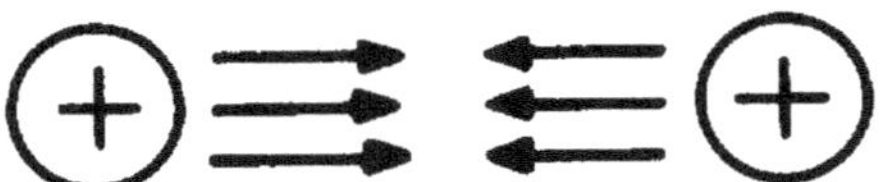

Figure 3-2.—Electrostatic field repels two like charged particles.

If two unlike charges are placed on opposite sides of an atom whose outermost electrons cannot escape their orbits, the orbits of the electrons are distorted as shown in figure 3-3. Figure 3-3(A) shows the normal orbit. Part (B) of the figure shows the same orbit in the presence of charged particles. Since the electron is a negative charge, the positive charge attracts the electrons, pulling the electrons closer to the positive charge. The negative charge repels the electrons, pushing them further from the negative charge. It is this ability of an electrostatic field to attract and to repel charges that allows the capacitor to store energy.

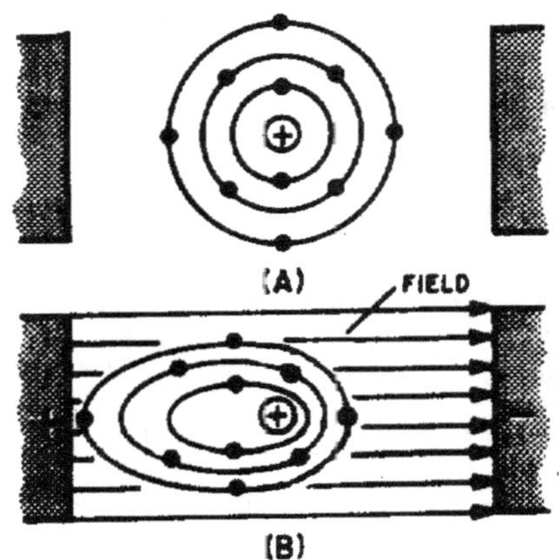

Figure 3-3.—Distortion of electron orbital paths due to electrostatic force.

Q2. State four characteristics of electrostatic lines of force.

Q3. An electron moves into the electrostatic field between a positive charge and a negative charge. Toward which charge will the electron move?

THE SIMPLE CAPACITOR

A simple capacitor consists of two metal plates separated by an insulating material called a dielectric, as illustrated in figure 3-4. Note that one plate is connected to the positive terminal of a battery; the other plate is connected through a closed switch (S1) to the negative terminal of the battery. Remember, an insulator is a material whose electrons cannot easily escape their orbits. Due to the battery voltage, plate A is charged positively and plate B is charged negatively. (How this happens is explained later in this chapter.) Thus an electrostatic field is set up between the positive and negative plates. The electrons on the negative plate (plate B) are attracted to the positive charges on the positive plate (plate A).

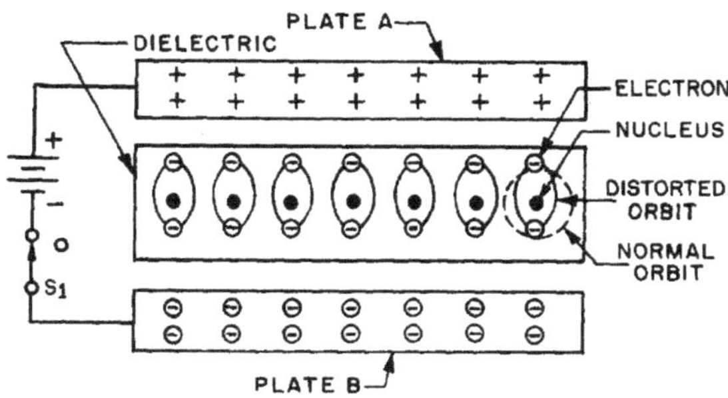

Figure 3-4.—Distortion of electron orbits in a dielectric.

Notice that the orbits of the electrons in the dielectric material are distorted by the electrostatic field. The distortion occurs because the electrons in the dielectric are attracted to the top plate while being repelled from the bottom plate. When switch S1 is opened, the battery is removed from the circuit and the charge is retained by the capacitor. This occurs because the dielectric material is an insulator, and the electrons in the bottom plate (negative charge) have no path to reach the top plate (positive charge). The distorted orbits of the atoms of the dielectric plus the electrostatic force of attraction between the two plates hold the positive and negative charges in their original position. Thus, the energy which came from the battery is now stored in the electrostatic field of the capacitor. Two slightly different symbols for representing a capacitor are shown in figure 3-5. Notice that each symbol is composed of two plates separated by a space that represents the dielectric. The curved plate in (B) of the figure indicates the plate should be connected to a negative polarity.

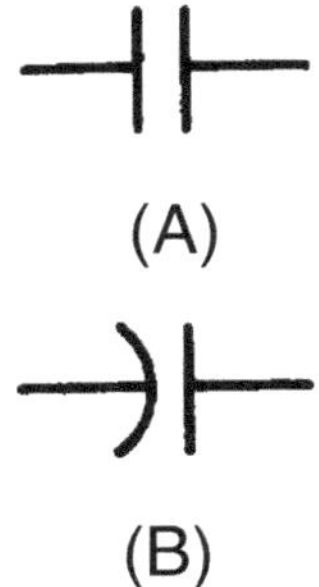

Figure 3-5.—Circuit symbols for capacitors.

Q4. What are the basic parts of a capacitor?

THE FARAD

Capacitance is measured in units called FARADS. A one-farad capacitor stores one coulomb (a unit of charge (Q) equal to 6.28×10^{18} electrons) of charge when a potential of 1 volt is applied across the terminals of the capacitor. This can be expressed by the formula:

$$C(\text{farads}) = \frac{Q\,(\text{coulombs})}{E\,(\text{volts})}$$

The farad is a very large unit of measurement of capacitance. For convenience, the microfarad (abbreviated μF) or the picofarad (abbreviated μF) is used. One (1.0) microfarad is equal to 0.000001 farad or 1×10^{-6} farad, and 1.0 picofarad is equal to 0.000000000001 farad or 1.0×10^{-12} farad. Capacitance is a physical property of the capacitor and does not depend on circuit characteristics of voltage, current, and resistance. A given capacitor always has the same value of capacitance (farads) in one circuit as in any other circuit in which it is connected.

Q5. Define the term "farad."

Q6. What is the mathematical relationship between a farad, a microfarad, and a picofarad.

FACTORS AFFECTING THE VALUE OF CAPACITANCE

The value of capacitance of a capacitor depends on three factors:

- The area of the plates.
- The distance between the plates.
- The dielectric constant of the material between the plates.

PLATE AREA affects the value of capacitance in the same manner that the size of a container affects the amount of water that can be held by the container. A capacitor with the large plate area can store more charges than a capacitor with a small plate area. Simply stated, "the larger the plate area, the larger the capacitance".

The second factor affecting capacitance is the DISTANCE BETWEEN THE PLATES. Electrostatic lines of force are strongest when the charged particles that create them are close together. When the charged particles are moved further apart, the lines of force weaken, and the ability to store a charge decreases.

The third factor affecting capacitance is the DIELECTRIC CONSTANT of the insulating material between the plates of a capacitor. The various insulating materials used as the dielectric in a capacitor differ in their ability to respond to (pass) electrostatic lines of force. A dielectric material, or insulator, is rated as to its ability to respond to electrostatic lines of force in terms of a figure called the DIELECTRIC CONSTANT. A dielectric material with a high dielectric constant is a better insulator than a dielectric material with a low dielectric constant. Dielectric constants for some common materials are given in the following list:

Material	Constant
Vacuum	1.0000
Air	1.0006
Paraffin paper	3.5
Glass	5 to 10
Mica	3 to 6
Rubber	2.5 to 35
Wood	2.5 to 8
Glycerine (15ºC)	56
Petroleum	2
Pure water	81

Notice the dielectric constant for a vacuum. Since a vacuum is the standard of reference, it is assigned a constant of one. The dielectric constants of all materials are compared to that of a vacuum. Since the dielectric constant of air has been determined to be approximately the same as that of a vacuum, the dielectric constant of AIR is also considered to be equal to one.

The formula used to compute the value of capacitance is:

$$C = 0.2249\left(\frac{KA}{d}\right)$$

Where C = capacitance in picofarads

A = area of one plate, in square inches

d = distance between the plates, in inches

K = dielectric constant of the insulating material

0.2249 = a constant resulting from conversion from Metric to English units.

Example: Find the capacitance of a parallel plate capacitor with paraffin paper as the dielectric.

Given: $K = 3.5$

$d = 0.05$ inch

$A = 12$ square inches

Solution: $C = 0.2249\left(\frac{KA}{d}\right)$

$$C = 0.2249\left(\frac{3.5 \times 12}{0.05}\right)$$

$C = 189$ picofarads

By examining the above formula you can see that capacitance varies directly as the dielectric constant and the area of the capacitor plates, and inversely as the distance between the plates.

Q7. State three factors that affect the capacitance of a capacitor.

Q8. A parallel plate capacitor has the following values: K = 81, d = .025 inches, A = 6 square inches. What is the capacitance of the capacitor?

VOLTAGE RATING OF CAPACITORS

In selecting or substituting a capacitor for use, consideration must be given to (1) the value of capacitance desired and (2) the amount of voltage to be applied across the capacitor. If the voltage applied across the capacitor is too great, the dielectric will break down and arcing will occur between the capacitor plates. When this happens the capacitor becomes a short-circuit and the flow of direct current through it can cause damage to other electronic parts. Each capacitor has a voltage rating (a working voltage) that should not be exceeded.

The working voltage of the capacitor is the maximum voltage that can be steadily applied without danger of breaking down the dielectric. The working voltage depends on the type of material used as the dielectric and on the thickness of the dialectic. (A high-voltage capacitor that has a thick dielectric must have a relatively large plate area in order to have the same capacitance as a similar low-voltage capacitor having a thin dielectric.) The working voltage also depends on the applied frequency because the losses, and the resultant heating effect, increase as the frequency increases.

A capacitor with a voltage rating of 500 volts dc cannot be safely subjected to an alternating voltage or a pulsating direct voltage having an effective value of 500 volts. Since an alternating voltage of 500 volts (rms) has a peak value of 707 volts, a capacitor to which it is applied should have a working voltage of at least 750 volts. In practice, a capacitor should be selected so that its working voltage is at least 50 percent greater than the highest effective voltage to be applied to it.

CAPACITOR LOSSES

Power loss in a capacitor may be attributed to dielectric hysteresis and dielectric leakage. Dielectric hysteresis may be defined as an effect in a dielectric material similar to the hysteresis found in a magnetic material. It is the result of changes in orientation of electron orbits in the dielectric because of the rapid reversals of the polarity of the line voltage. The amount of power loss due to dielectric hysteresis depends upon the type of dielectric used. A vacuum dielectric has the smallest power loss.

Dielectric leakage occurs in a capacitor as the result of LEAKAGE CURRENT through the dielectric. Normally it is assumed that the dielectric will effectively prevent the flow of current through the capacitor. Although the resistance of the dielectric is extremely high, a minute amount of current does flow. Ordinarily this current is so small that for all practical purposes it is ignored. However, if the leakage through the dielectric is abnormally high, there will be a rapid loss of charge and an overheating of the capacitor.

The power loss of a capacitor is determined by loss in the dielectric. If the loss is negligible and the capacitor returns the total charge to the circuit, it is considered to be a perfect capacitor with a power loss of zero.

Q9. Name two types of power losses associated with a capacitor.

Q10.

a. Define the term "working voltage" of a capacitor.

b. What should be the working voltage of a capacitor in a circuit that is operating at 600 volts?

CHARGING AND DISCHARGING A CAPACITOR

CHARGING

In order to better understand the action of a capacitor in conjunction with other components, the charge and discharge actions of a purely capacitive circuit are analyzed first. For ease of explanation the capacitor and voltage source shown in figure 3-6 are assumed to be perfect (no internal resistance), although this is impossible in practice.

In figure 3-6(A), an uncharged capacitor is shown connected to a four-position switch. With the switch in position 1 the circuit is open and no voltage is applied to the capacitor. Initially each plate of the capacitor is a neutral body and until a difference of potential is impressed across the capacitor, no electrostatic field can exist between the plates.

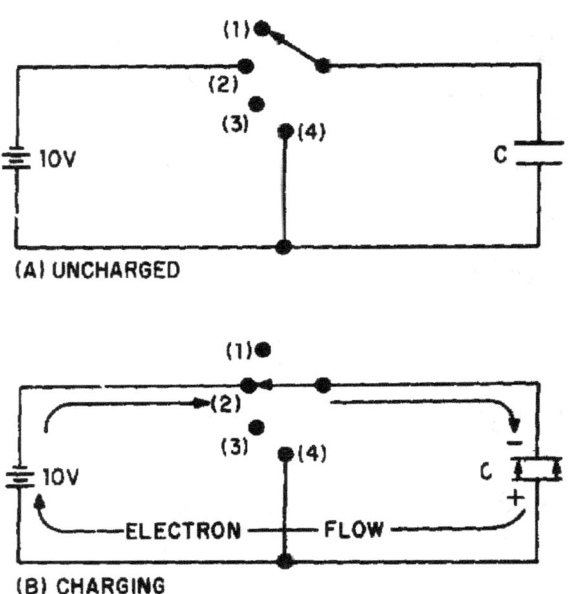

Figure 3-6.—Charging a capacitor.

To CHARGE the capacitor, the switch must be thrown to position 2, which places the capacitor across the terminals of the battery. Under the assumed perfect conditions, the capacitor would reach full charge instantaneously. However, the charging action is spread out over a period of time in the following discussion so that a step-by-step analysis can be made.

At the instant the switch is thrown to position 2 (fig. 3-6(B)), a displacement of electrons occurs simultaneously in all parts of the circuit. This electron displacement is directed away from the negative terminal and toward the positive terminal of the source (the battery). A brief surge of current will flow as the capacitor charges.

If it were possible to analyze the motion of the individual electrons in this surge of charging current, the following action would be observed. See figure 3-7.

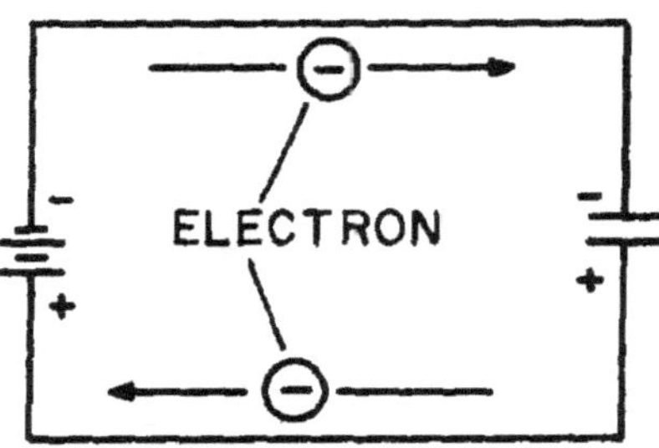

Figure 3-7.—Electron motion during charge.

At the instant the switch is closed, the positive terminal of the battery extracts an electron from the bottom conductor. The negative terminal of the battery forces an electron into the top conductor. At this same instant an electron is forced into the top plate of the capacitor and another is pulled from the bottom plate. Thus, in every part of the circuit a clockwise DISPLACEMENT of electrons occurs simultaneously.

As electrons accumulate on the top plate of the capacitor and others depart from the bottom plate, a difference of potential develops across the capacitor. Each electron forced onto the top plate makes that plate more negative, while each electron removed from the bottom causes the bottom plate to become more positive. Notice that the polarity of the voltage which builds up across the capacitor is such as to oppose the source voltage. The source voltage (emf) forces current around the circuit of figure 3-7 in a clockwise direction. The emf developed across the capacitor, however, has a tendency to force the current in a counterclockwise direction, opposing the source emf. As the capacitor continues to charge, the voltage across the capacitor rises until it is equal to the source voltage. Once the capacitor voltage equals the source voltage, the two voltages balance one another and current ceases to flow in the circuit.

In studying the charging process of a capacitor, you must be aware that NO current flows THROUGH the capacitor. The material between the plates of the capacitor must be an insulator. However, to an observer stationed at the source or along one of the circuit conductors, the action has all the appearances of a true flow of current, even though the insulating material between the plates of the capacitor prevents the current from having a complete path. The current which appears to flow through a capacitor is called DISPLACEMENT CURRENT.

When a capacitor is fully charged and the source voltage is equaled by the counter electromotive force (cemf) across the capacitor, the electrostatic field between the plates of the capacitor is maximum. (Look again at figure 3-4.) Since the electrostatic field is maximum the energy stored in the dielectric is also maximum.

If the switch is now opened as shown in figure 3-8(A), the electrons on the upper plate are isolated. The electrons on the top plate are attracted to the charged bottom plate. Because the dielectric is an insulator, the electrons can not cross the dielectric to the bottom plate. The charges on both plates will be effectively trapped by the electrostatic field and the capacitor will remain charged indefinitely. You should note at this point that the insulating dielectric material in a practical capacitor is not perfect and small leakage current will flow through the dielectric. This current will eventually dissipate the charge. However, a high quality capacitor may hold its charge for a month or more.

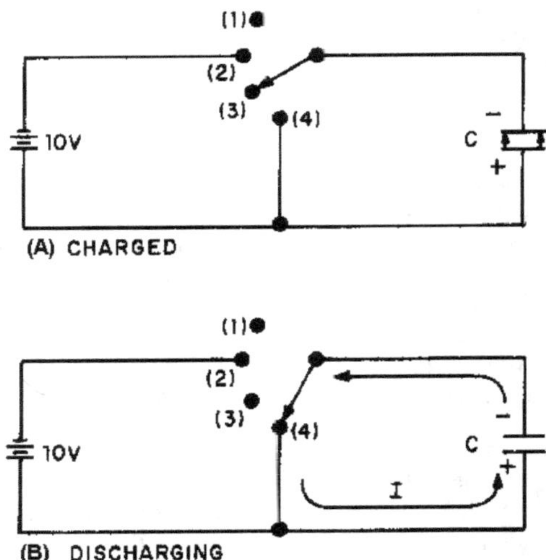

Figure 3-8.—Discharging a capacitor.

To review briefly, when a capacitor is connected across a voltage source, a surge of charging current flows. This charging current develops a cemf across the capacitor which opposes the applied voltage. When the capacitor is fully charged, the cemf is equal to the applied voltage and charging current ceases. At full charge, the electrostatic field between the plates is at maximum intensity and the energy stored in the dielectric is maximum. If the charged capacitor is disconnected from the source, the charge will be retained for some period of time. The length of time the charge is retained depends on the amount of leakage current present. Since electrical energy is stored in the capacitor, a charged capacitor can act as a source emf.

DISCHARGING

To DISCHARGE a capacitor, the charges on the two plates must be neutralized. This is accomplished by providing a conducting path between the two plates as shown in figure 3-8(B). With the switch in position (4) the excess electrons on the negative plate can flow to the positive plate and neutralize its charge. When the capacitor is discharged, the distorted orbits of the electrons in the dielectric return to their normal positions and the stored energy is returned to the circuit. It is important for you to note that a capacitor does not consume power. The energy the capacitor draws from the source is recovered when the capacitor is discharged.

Q11. State what happens to the electrons in a capacitor circuit when (a) the capacitor is charging and (b) the capacitor is discharging.

CHARGE AND DISCHARGE OF AN RC SERIES CIRCUIT

Ohm's law states that the voltage across a resistance is equal to the current through the resistance times the value of the resistance. This means that a voltage is developed across a resistance ONLY WHEN CURRENT FLOWS through the resistance.

A capacitor is capable of storing or holding a charge of electrons. When uncharged, both plates of the capacitor contain essentially the same number of free electrons. When charged, one plate contains more free electrons than the other plate. The difference in the number of electrons is a measure of the charge on the capacitor. The accumulation of this charge builds up a voltage across the terminals of the capacitor, and the charge continues to increase until this voltage equals the applied voltage. The charge in a capacitor is related to the capacitance and voltage as follows:

$$Q = CE,$$

in which Q is the charge in coulombs, C the capacitance in farads, and E the emf across the capacitor in volts.

CHARGE CYCLE

A voltage divider containing resistance and capacitance is connected in a circuit by means of a switch, as shown at the top of figure 3-9. Such a series arrangement is called an RC series circuit.

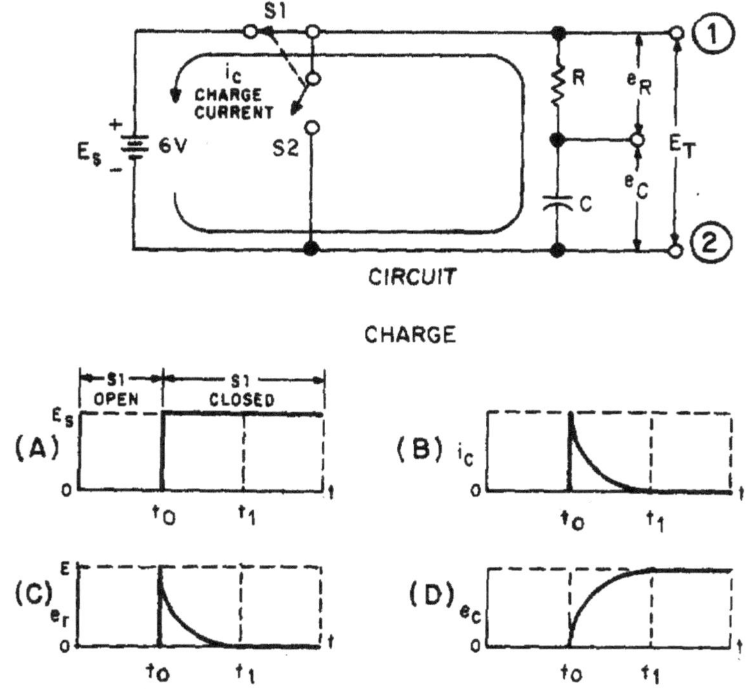

Figure 3-9.—Charge of an RC series circuit.

In explaining the charge and discharge cycles of an RC series circuit, the time interval from time t_0 (time zero, when the switch is first closed) to time t_1 (time one, when the capacitor reaches full charge or discharge potential) will be used. (Note that switches S1 and S2 move at the same time and can never both be closed at the same time.)

When switch S1 of the circuit in figure 3-9 is closed at t_0, the source voltage (E_S) is instantly felt across the entire circuit. Graph (A) of the figure shows an instantaneous rise at time t_0 from zero to source voltage (E_S = 6 volts). The total voltage can be measured across the circuit between points 1 and 2. Now look at graph (B) which represents the charging current in the capacitor (i_c). At time t_0, charging current is MAXIMUM. As time elapses toward time t_1, there is a continuous decrease in current flowing into the capacitor. The decreasing flow is caused by the voltage buildup across the capacitor. At time t_1, current flowing in the capacitor stops. At this time, the capacitor has reached full charge and has stored maximum energy in its electrostatic field. Graph (C) represents the voltage drop (e) across the resistor (R). The value of e_r is determined by the amount of current flowing through the resistor on its way to the capacitor. At time t_0 the current flowing to the capacitor is maximum. Thus, the voltage drop across the resistor is maximum (E = IR). As time progresses toward time t_1, the current flowing to the capacitor steadily decreases and causes the voltage developed across the resistor (R) to steadily decrease. When time t_1 is reached, current flowing to the capacitor is stopped and the voltage developed across the resistor has decreased to zero.

You should remember that capacitance opposes a change in voltage. This is shown by comparing graph (A) to graph (D). In graph (A) the voltage changed instantly from 0 volts to 6 volts across the circuit, while the voltage developed across the capacitor in graph (D) took the entire time interval from time t_o to time t_1 to reach 6 volts. The reason for this is that in the first instant at time t_0, maximum current flows through R and the entire circuit voltage is dropped across the resistor. The voltage impressed across the capacitor at t_0 is zero volts. As time progresses toward t_1, the decreasing current causes progressively less voltage to be dropped across the resistor (R), and more voltage builds up across the capacitor (C). At time t_1, the voltage felt across the capacitor is equal to the source voltage (6 volts), and the voltage dropped across the resistor (R) is equal to zero. This is the complete charge cycle of the capacitor.

As you may have noticed, the processes which take place in the time interval t_0 to t_1 in a series RC circuit are exactly opposite to those in a series LR circuit.

For your comparison, the important points of the charge cycle of RC and LR circuits are summarized in table 3-1.

Table 3-1.—Summary of Capacitive and Inductive Characteristics

		TIME ZERO (t_0)	TIME BETWEEN t_0 AND t_1	TIME ONE (t_1)
CIRCUIT CURRENT		MAXIMUM	DECREASING	ZERO
		ZERO	INCREASING	MAXIMUM
VOLTAGE DEVELOPED ACROSS THE RESISTOR		MAXIMUM	DECREASING	ZERO
		ZERO	INCREASING	MAXIMUM
VOLTAGE DEVELOPED ACROSS CAPACITOR/ INDUCTOR		ZERO	INCREASING	MAXIMUM
		MAXIMUM	DECREASING	ZERO

DISCHARGE CYCLE

In figure 3-10 at time t_0, the capacitor is fully charged. When S1 is open and S2 closes, the capacitor discharge cycle starts. At the first instant, circuit voltage attempts to go from source potential (6 volts) to zero volts, as shown in graph (A). Remember, though, the capacitor during the charge cycle has stored energy in an electrostatic field.

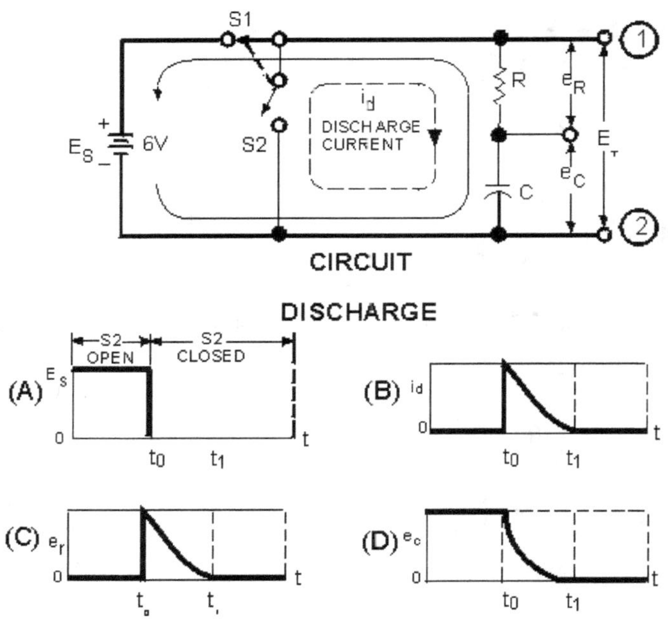

Figure 3-10.—Discharge of an RC Series circuit.

Because S2 is closed at the same time S1 is open, the stored energy of the capacitor now has a path for current to flow. At t_0, discharge current (i_d) from the bottom plate of the capacitor through the resistor (R) to the top plate of the capacitor (C) is maximum. As time progresses toward t_1, the discharge current steadily decreases until at time t_1 it reaches zero, as shown in graph (B).

The discharge causes a corresponding voltage drop across the resistor as shown in graph (C). At time t_0, the current through the resistor is maximum and the voltage drop (e_r) across the resistor is maximum. As the current through the resistor decreases, the voltage drop across the resistor decreases until at t_1 it has reached a value of zero. Graph (D) shows the voltage across the capacitor (e_c) during the discharge cycle. At time t_0 the voltage is maximum and as time progresses toward time t_1, the energy stored in the capacitor is depleted. At the same time the voltage across the resistor is decreasing, the voltage (e) across the capacitor is decreasing until at time t_1 the voltage (e_c) reaches zero.

By comparing graph (A) with graph (D) of figure 3-10, you can see the effect that capacitance has on a change in voltage. If the circuit had not contained a capacitor, the voltage would have ceased at the instant S1 was opened at time t_0. Because the capacitor is in the circuit, voltage is applied to the circuit until the capacitor has discharged completely at t_1. The effect of capacitance has been to oppose this change in voltage.

Q12. At what instant does the greatest voltage appear across the resistor in a series RC circuit when the capacitor is charging?

Q13. What is the voltage drop across the resistor in an RC charging circuit when the charge on the capacitor is equal to the battery voltage?

RC TIME CONSTANT

The time required to charge a capacitor to 63 percent (actually 63.2 percent) of full charge or to discharge it to 37 percent (actually 36.8 percent) of its initial voltage is known as the TIME CONSTANT (TC) of the circuit. The charge and discharge curves of a capacitor are shown in figure 3-11. Note that the charge curve is like the curve in figure 3-9, graph (D), and the discharge curve like the curve in figure 3-9, graph (B).

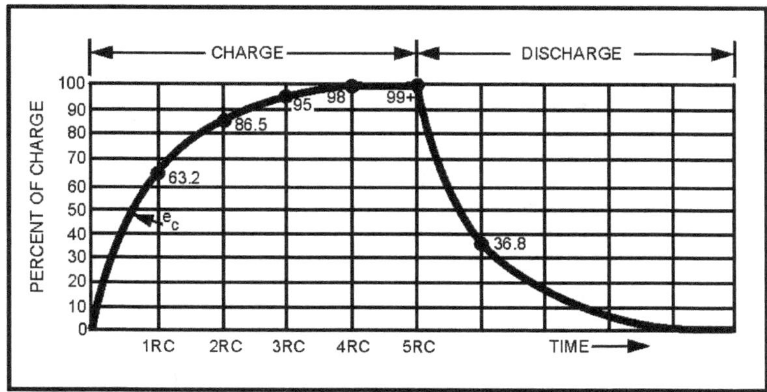

Figure 3-11.—RC time constant.

The value of the time constant in seconds is equal to the product of the circuit resistance in ohms and the circuit capacitance in farads. The value of one time constant is expressed mathematically as t = RC. Some forms of this formula used in calculating RC time constants are:

t (in seconds) = R (in ohms) × C (in farads)

t (in seconds) = R (in megohms) × C (in microfarads)

t (in microseconds) = R (in ohms) × C (in microfarads)

t (in microseconds) = R (in megohms) × C (in picofarads)

Q14. What is the RC time constant of a series RC circuit that contains a 12-megohm resistor and a 12-microfarad capacitor?

UNIVERSAL TIME CONSTANT CHART

Because the impressed voltage and the values of R and C or R and L in a circuit are usually known, a UNIVERSAL TIME CONSTANT CHART (fig. 3-12) can be used to find the time constant of the circuit. Curve A is a plot of both capacitor voltage during charge and inductor current during growth. Curve B is a plot of both capacitor voltage during discharge and inductor current during decay.

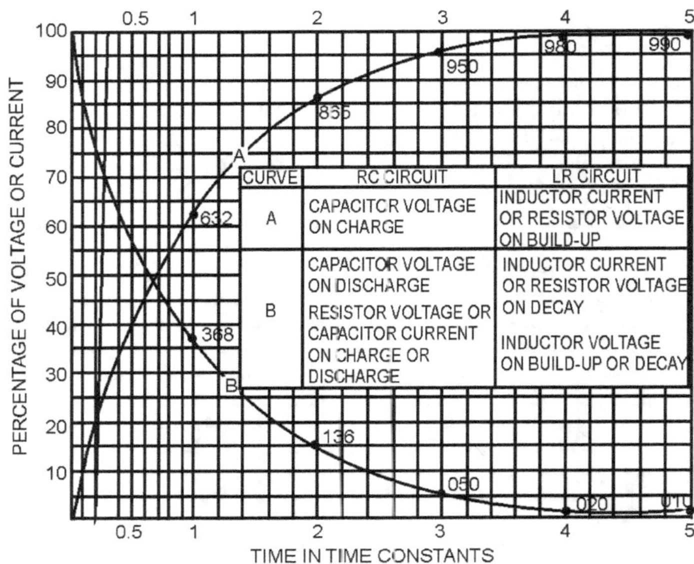

Figure 3-12.—Universal time constant chart for RC and RL circuit.

The time scale (horizontal scale) is graduated in terms of the RC or L/R time constants so that the curves may be used for any value of R and C or L and R. The voltage and current scales (vertical scales) are graduated in terms of percentage of the maximum voltage or current so that the curves may be used for any value of voltage or current. If the time constant and the initial or final voltage for the circuit in question are known, the voltages across the various parts of the circuit can be obtained from the curves for any time after the switch is closed, either on charge or discharge. The same reasoning is true of the current in the circuit.

The following problem illustrates how the universal time constant chart may be used.

An RC circuit is to be designed in which a capacitor (C) must charge to 20 percent (0.20) of the maximum charging voltage in 100 microseconds (0.0001 second). Because of other considerations, the resistor (R) must have a value of 20,000 ohms. What value of capacitance is needed?

Given: Percent of charge = 20% (.20)

$t = 100 \mu s$

$R = 20{,}000\ \Omega$

Find: The capacitance of capacitor C.

Solution: Because the only values given are in units of time and resistance, a variation of the formula to find RC time is used:

$$RC = R \times C$$

where: 1 RC time constant $= R \times C$
and R is known.

Transpose the formula to:

$$C = \frac{RC}{R}$$

Find the value of RC by referring to the universal time constant chart in figure 3-12 and proceed as follows:

- Locate the 20 point on the vertical scale at the left side of the chart (percentage).
- Follow the horizontal line from this point to intersect curve A.
- Follow an imaginary vertical line from the point of intersection on curve A downward to cross the RC scale at the bottom of the chart.

Note that the vertical line crosses the horizontal scale at about .22 RC as illustrated below:

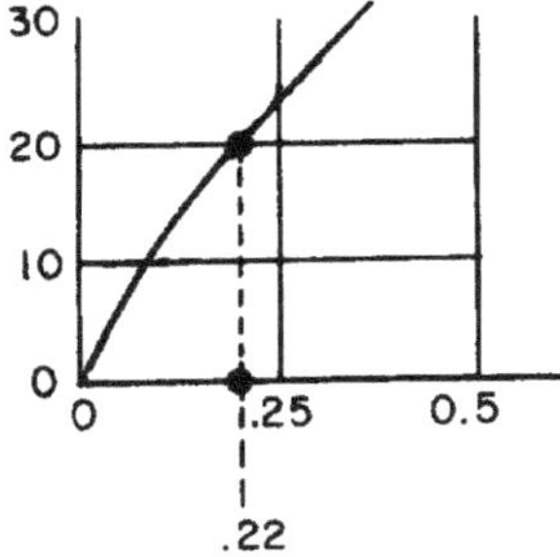

The value selected from the graph means that a capacitor (including the one you are solving for) will reach twenty percent of full charge in twenty-two one hundredths (.22) of one RC time constant. Remember that it takes 100 μs for the capacitor to reach 20% of full charge. Since 100 μs is equal to .22 RC (twenty-two one-hundredths), then the time required to reach one RC time constant must be equal to:

$$.22\,RC = 100\ \mu s$$

$$RC = \frac{1}{.22} \times 100\ \mu s$$

$$RC = \frac{100\ \mu s}{.22}$$

$$RC = 454.54\ \mu s \text{ (rounded off to } 455\ \mu s)$$

$$RC = 455\ \mu s$$

Now use the following formula to find C:

$$C = \frac{RC}{R}$$

$$C = \frac{455\ \mu s}{20{,}000\ \text{ohms}}$$

$$C = 0.0227\ \mu F$$

$$C = .023\ \mu F$$

To summarize the above procedures, the problem and solution are shown below without the step by step explanation.

$$\text{Given: Percent of charge} = 20\%\ (.20)$$

$$t = 100\ \mu s$$

$$R = 20{,}000\ \text{ohms}$$

Transpose the RC time constant formula as follows:

$$R \times C = RC$$

$$C = \frac{RC}{R}$$

$$\text{Find: } RC$$

$$.22\,RC = 100\ \mu s$$

$$RC = \frac{100\mu s}{.22}$$

$$RC = 455\ \mu s$$

Substitute the R and RC values into the formula:

$$C = \frac{RC}{R}$$

$$C = \frac{455\ \mu s}{20{,}000}\ \text{ohms}$$

$$C = .023\ \mu s$$

The graphs shown in figure 3-11 and 3-12 are not entirely complete. That is, the charge or discharge (or the growth or decay) is not quite complete in 5 RC or 5 L/R time constants. However, when the values reach 0.99 of the maximum (corresponding to 5 RC or 5 L/R), the graphs may be considered accurate enough for all practical purposes.

Q15. A circuit is to be designed in which a capacitor must charge to 40 percent of the maximum charging voltage in 200 microseconds. The resistor to be used has a resistance of 40,000 ohms. What size capacitor must be used? (Use the universal time constant chart in figure 3-12.)

CAPACITORS IN SERIES AND PARALLEL

Capacitors may be connected in series or in parallel to obtain a resultant value which may be either the sum of the individual values (in parallel) or a value less than that of the smallest capacitance (in series).

CAPACITORS IN SERIES

The overall effect of connecting capacitors in series is to move the plates of the capacitors further apart. This is shown in figure 3-13. Notice that the junction between C1 and C2 has both a negative and a positive charge. This causes the junction to be essentially neutral. The total capacitance of the circuit is developed between the left plate of C1 and the right plate of C2. Because these plates are farther apart, the total value of the capacitance in the circuit is decreased. Solving for the total capacitance (C_T) of capacitors connected in series is similar to solving for the total resistance (R_T) of resistors connected in parallel.

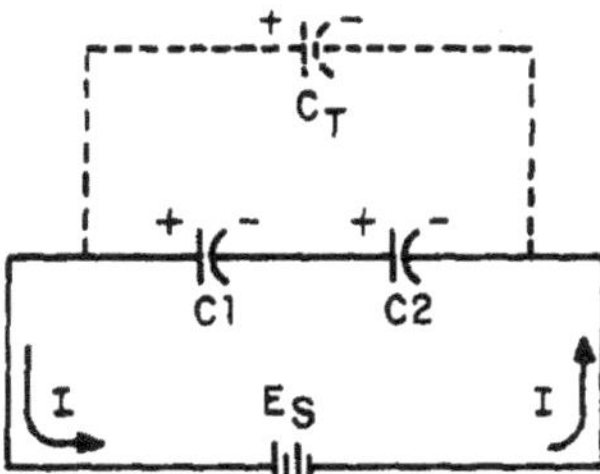

Figure 3-13.—Capacitors in series.

Note the similarity between the formulas for R_T and C_T:

$$R_T = \frac{1}{\frac{1}{R1} + \frac{1}{R2} + \dots \frac{1}{R_n}}$$

$$C_T = \frac{1}{\frac{1}{C1} + \frac{1}{C2} + \dots \frac{1}{C_n}}$$

If the circuit contains more than two capacitors, use the above formula. If the circuit contains only two capacitors, use the below formula:

$$C_T = \frac{C1 \times C2}{C1 + C2}$$

Note: All values for C_T, C1, C2, C3,... C_n should be in farads. It should be evident from the above formulas that the total capacitance of capacitors in series is less than the capacitance of any of the individual capacitors.

Example: Determine the total capacitance of a series circuit containing three capacitors whose values are 0.01 μF, 0.25 μF, and 50,000 pF, respectively.

Given: $C1 = 0.01\,\mu s$

$C2 = 0.25\,\mu s$

$C3 = 50{,}000\,pF$

Solution:

$$C_T = \frac{1}{\frac{1}{C1}+\frac{1}{C2}+\frac{1}{C3}}$$

$$C_T = \frac{1}{\frac{1}{.01\,\mu F}+\frac{1}{.25\,\mu F}+\frac{1}{50{,}000\,pF}}$$

$$C_T = \frac{1}{\frac{1}{1\times 10^{-8}}+\frac{1}{25\times 10^{-8}}+\frac{1}{5\times 10^{-8}}}F$$

$$C_T = \frac{1}{100\times 10^{6}+4\times 10^{6}+20\times 10^{6}}F$$

$$C_T = \frac{1}{124\times 10^{6}}F$$

$$C_T = 0.008\,\mu F$$

The total capacitance of 0.008μF is slightly smaller than the smallest capacitor (0.01μF).

CAPACITORS IN PARALLEL

When capacitors are connected in parallel, one plate of each capacitor is connected directly to one terminal of the source, while the other plate of each capacitor is connected to the other terminal of the source. Figure 3-14 shows all the negative plates of the capacitors connected together, and all the positive plates connected together. C_T, therefore, appears as a capacitor with a plate area equal to the sum of all the individual plate areas. As previously mentioned, capacitance is a direct function of plate area. Connecting capacitors in parallel effectively increases plate area and thereby increases total capacitance.

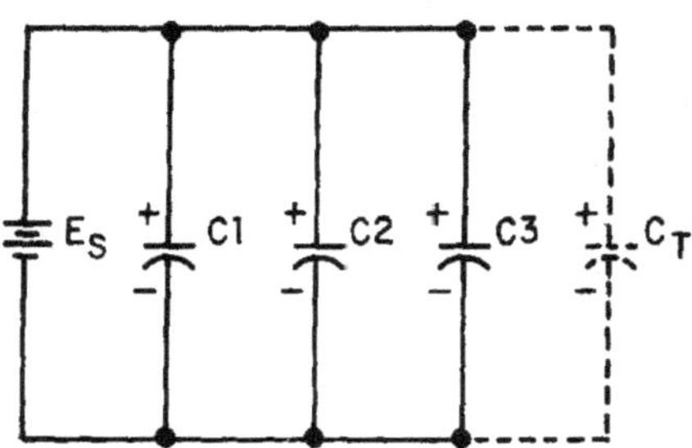

Figure 3-14.—Parallel capacitive circuit.

For capacitors connected in parallel the total capacitance is the sum of all the individual capacitances. The total capacitance of the circuit may by calculated using the formula:

$$C_T = C1 + C2 + C3 + \ldots\ldots C_n$$

where all capacitances are in the same units.

Example: Determine the total capacitance in a parallel capacitive circuit containing three capacitors whose values are 0.03 μF, 2.0 μF, and 0.25 μF, respectively.

Given: $C1 = 0.03\,\mu F$

$C2 = 2\,\mu F$

$C3 = 0.25\,\mu F$

Solution: $C_T = C1 + C2 + C3$

$C_T = 0.03\,\mu F + 2.0\,\mu F + 0.25\,\mu F$

$C_T = 2.28\,\mu F$

Q16. What is the total capacitance of a circuit that contains two capacitors (10 μF and 0.1 μF) wired together in series?

Q17. What is the total capacitance of a circuit in which four capacitors (10 μF, 21 μF, 0.1 μF and 2 μF) are connected in parallel?

FIXED CAPACITOR

A fixed capacitor is constructed in such manner that it possesses a fixed value of capacitance which cannot be adjusted. A fixed capacitor is classified according to the type of material used as its dielectric, such as paper, oil, mica, or electrolyte.

A PAPER CAPACITOR is made of flat thin strips of metal foil conductors that are separated by waxed paper (the dielectric material). Paper capacitors usually range in value from about 300 picofarads to about 4 microfarads. The working voltage of a paper capacitor rarely exceeds 600 volts. Paper capacitors are sealed with wax to prevent the harmful effects of moisture and to prevent corrosion and leakage.

Many different kinds of outer covering are used on paper capacitors, the simplest being a tubular cardboard covering. Some types of paper capacitors are encased in very hard plastic. These types are very rugged and can be used over a much wider temperature range than can the tubular cardboard type. Figure 3-15(A) shows the construction of a tubular paper capacitor; part 3-15(B) shows a completed cardboard-encased capacitor.

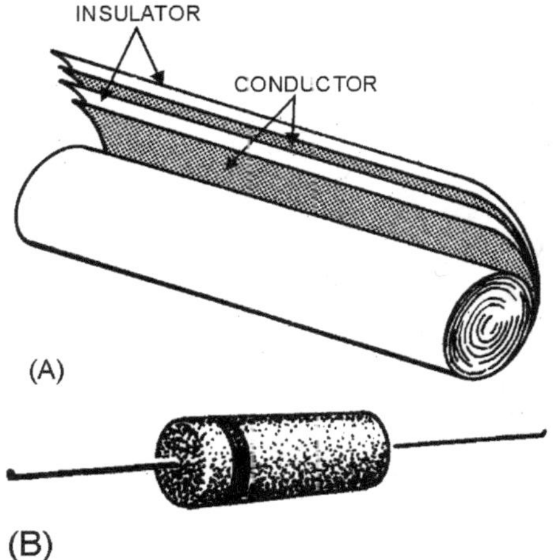

Figure 3-15.—Paper capacitor.

A MICA CAPACITOR is made of metal foil plates that are separated by sheets of mica (the dielectric). The whole assembly is encased in molded plastic. Figure 3-16(A) shows a cut-away view of a mica capacitor. Because the capacitor parts are molded into a plastic case, corrosion and damage to the plates and dielectric are prevented. In addition, the molded plastic case makes the capacitor mechanically stronger. Various types of terminals are used on mica capacitors to connect them into circuits. These terminals are also molded into the plastic case.

Mica is an excellent dielectric and can withstand a higher voltage than can a paper dielectric of the same thickness. Common values of mica capacitors range from approximately 50 picofarads to 0.02 microfarad. Some different shapes of mica capacitors are shown in figure 3-16(B).

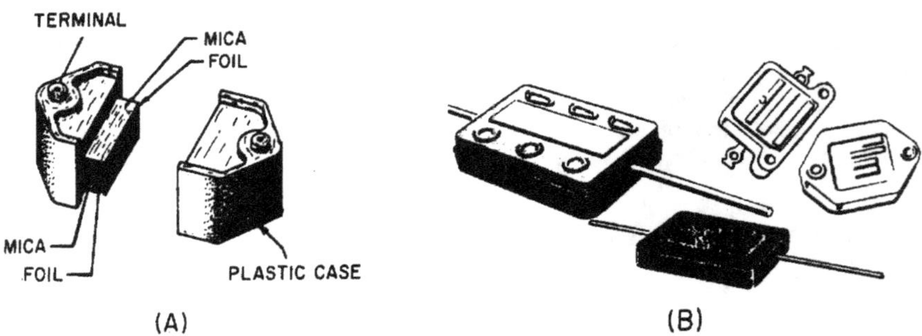

Figure 3-16.—Typical mica capacitors.

A CERAMIC CAPACITOR is so named because it contains a ceramic dielectric. One type of ceramic capacitor uses a hollow ceramic cylinder as both the form on which to construct the capacitor and as the dielectric material. The plates consist of thin films of metal deposited on the ceramic cylinder.

A second type of ceramic capacitor is manufactured in the shape of a disk. After leads are attached to each side of the capacitor, the capacitor is completely covered with an insulating moisture-proof coating. Ceramic capacitors usually range in value from 1 picofarad to 0.01 microfarad and may be used with voltages as high as 30,000 volts. Some different shapes of ceramic capacitors are shown in figure 3-17.

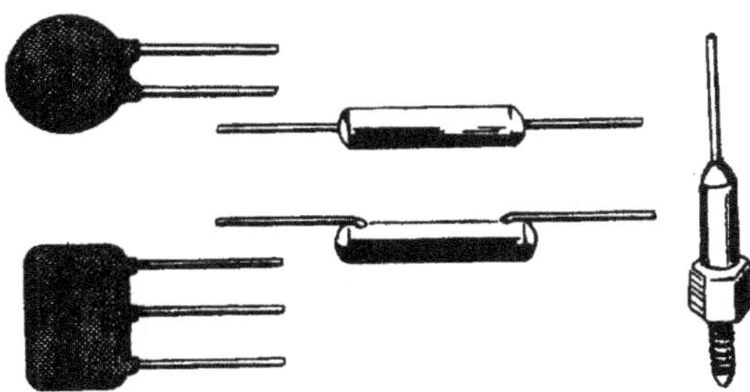

Figure 3-17.—Ceramic capacitors.

Examples of ceramic capacitors.

An ELECTROLYTIC CAPACITOR is used where a large amount of capacitance is required. As the name implies, an electrolytic capacitor contains an electrolyte. This electrolyte can be in the form of a liquid (wet electrolytic capacitor). The wet electrolytic capacitor is no longer in popular use due to the care needed to prevent spilling of the electrolyte.

A dry electrolytic capacitor consists essentially of two metal plates separated by the electrolyte. In most cases the capacitor is housed in a cylindrical aluminum container which acts as the negative terminal of the capacitor (see fig. 3-18). The positive terminal (or terminals if the capacitor is of the multisection type) is a lug (or lugs) on the bottom end of the container. The capacitance value(s) and the voltage rating of the capacitor are generally printed on the side of the aluminum case.

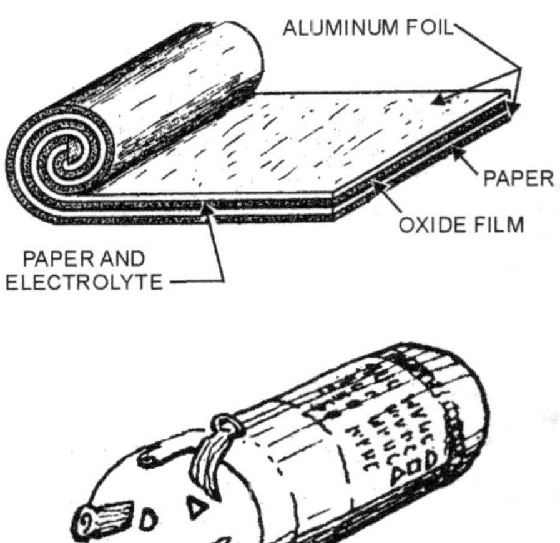

Figure 3-18.—Construction of an electrolytic capacitor.

An example of a multisection electrolytic capacitor is illustrated in figure 3-18(B). The four lugs at the end of the cylindrical aluminum container indicates that four electrolytic capacitors are enclosed in the can. Each section of the capacitor is electrically independent of the other sections. It is possible for one section to be defective while the other sections are still good. The can is the common negative connection to the four capacitors. Separate terminals are provided for the positive plates of the capacitors. Each capacitor is identified by an embossed mark adjacent to the lugs, as shown in figure 3-18(B). Note the identifying marks used on the electrolytic capacitor are the half moon, the triangle, the square, and no embossed mark. By looking at the bottom of the container and the identifying sheet pasted to the side of the container, you can easily identify the value of each section.

Internally, the electrolytic capacitor is constructed similarly to the paper capacitor. The positive plate consists of aluminum foil covered with an extremely thin film of oxide. This thin oxide film (which is formed by an electrochemical process) acts as the dielectric of the capacitor. Next to and in contact with the oxide is a strip of paper or gauze which has been impregnated with a paste-like electrolyte. The electrolyte acts as the negative plate of the capacitor. A second strip of aluminum foil is then placed against the electrolyte to provide electrical contact to the negative electrode (the electrolyte). When the three layers are in place they are rolled up into a cylinder as shown in figure 3-18(A).

An electrolytic capacitor has two primary disadvantages compared to a paper capacitor in that the electrolytic type is POLARIZED and has a LOW LEAKAGE RESISTANCE. This means that should the positive plate be accidentally connected to the negative terminal of the source, the thin oxide film dielectric will dissolve and the capacitor will become a conductor (i.e., it will short). The polarity of the terminals is normally marked on the case of the capacitor. Since an electrolytic capacitor is polarity sensitive, its use is ordinarily restricted to a dc circuit or to a circuit where a small ac voltage is superimposed on a dc voltage. Special electrolytic capacitors are available for certain ac applications, such as a motor starting capacitor. Dry electrolytic capacitors vary in size from about 4 microfarads to several thousand microfarads and have a working voltage of approximately 500 volts.

The type of dielectric used and its thickness govern the amount of voltage that can safely be applied to the electrolytic capacitor. If the voltage applied to the capacitor is high enough to cause the atoms of the dielectric material to become ionized, arcing between the plates will occur. In most other types of capacitors, arcing will destroy the capacitor. However, an electrolytic capacitor has the ability to be self-healing. If the arcing is small, the electrolytic will regenerate itself. If the arcing is too large, the capacitor will not self-heal and will become defective.

OIL CAPACITORS are often used in high-power electronic equipment. An oil-filled capacitor is nothing more than a paper capacitor that is immersed in oil. Since oil impregnated paper has a high dielectric constant, it can be used in the production of capacitors having a high capacitance value. Many capacitors will use oil with another dielectric material to prevent arcing between the plates. If arcing should occur between the plates of an oil-filled capacitor, the oil will tend to reseal the hole caused by the arcing. Such a capacitor is referred to as a SELF-HEALING capacitor.

VARIABLE CAPACITOR

A variable capacitor is constructed in such manner that its value of capacitance can be varied. A typical variable capacitor (adjustable capacitor) is the rotor-stator type. It consists of two sets of metal plates arranged so that the rotor plates move between the stator plates. Air is the dielectric. As the position of the rotor is changed, the capacitance value is likewise changed. This type of capacitor is used for tuning most radio receivers. Its physical appearance and its symbol are shown in figure 3-19.

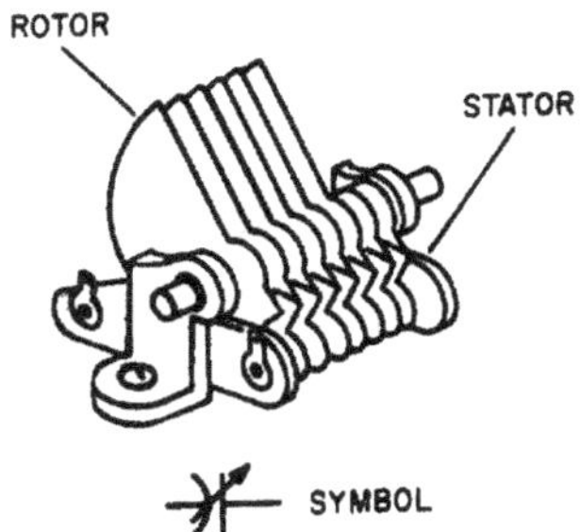

Figure 3-19.—Rotor-stator type variable capacitor.

Another type of variable capacitor (trimmer capacitor) and its symbol are shown in figure 3-20. This capacitor consists of two plates separated by a sheet of mica. A screw adjustment is used to vary the distance between the plates, thereby changing the capacitance.

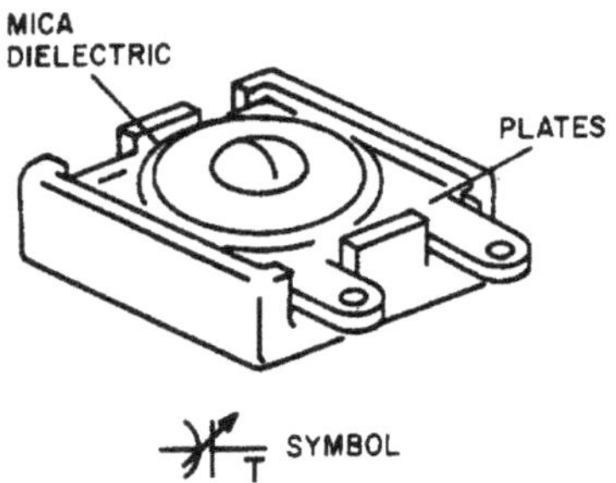

Figure 3-20.—Trimmer capacitor.

Q18.

a. An oxide-film dielectric is used in what type of capacitor?

b. A screw adjustment is used to vary the distance between the plates of what type of capacitor?

COLOR CODES FOR CAPACITORS

Although the capacitance value may be printed on the body of a capacitor, it may also be indicated by a color code. The color code used to represent capacitance values is similar to that used to represent resistance values. The color codes currently in use are the Joint Army-Navy (JAN) code and the Radio Manufacturers' Association (RMA) code.

For each of these codes, colored dots or bands are used to indicate the value of the capacitor. A mica capacitor, it should be noted, may be marked with either three dots or six dots. Both the three- and the six-dot codes are similar, but the six-dot code contains more information about electrical ratings of the capacitor, such as working voltage and temperature coefficient.

The capacitor shown in figure 3-21 represents either a mica capacitor or a molded paper capacitor. To determine the type and value of the capacitor, hold the capacitor so that the three arrows point left to right (>). The first dot at the base of the arrow sequence (the left-most dot) represents the capacitor TYPE. This dot is either black, white, silver, or the same color as the capacitor body. Mica is represented by a black or white dot and paper by a silver dot or dot having the same color as the body of the capacitor. The two dots to the immediate right of the type dot indicate the first and second digits of the capacitance value. The dot at the bottom right represents the multiplier to be used. The multiplier represents picofarads. The dot in the bottom center indicates the tolerance value of the capacitor.

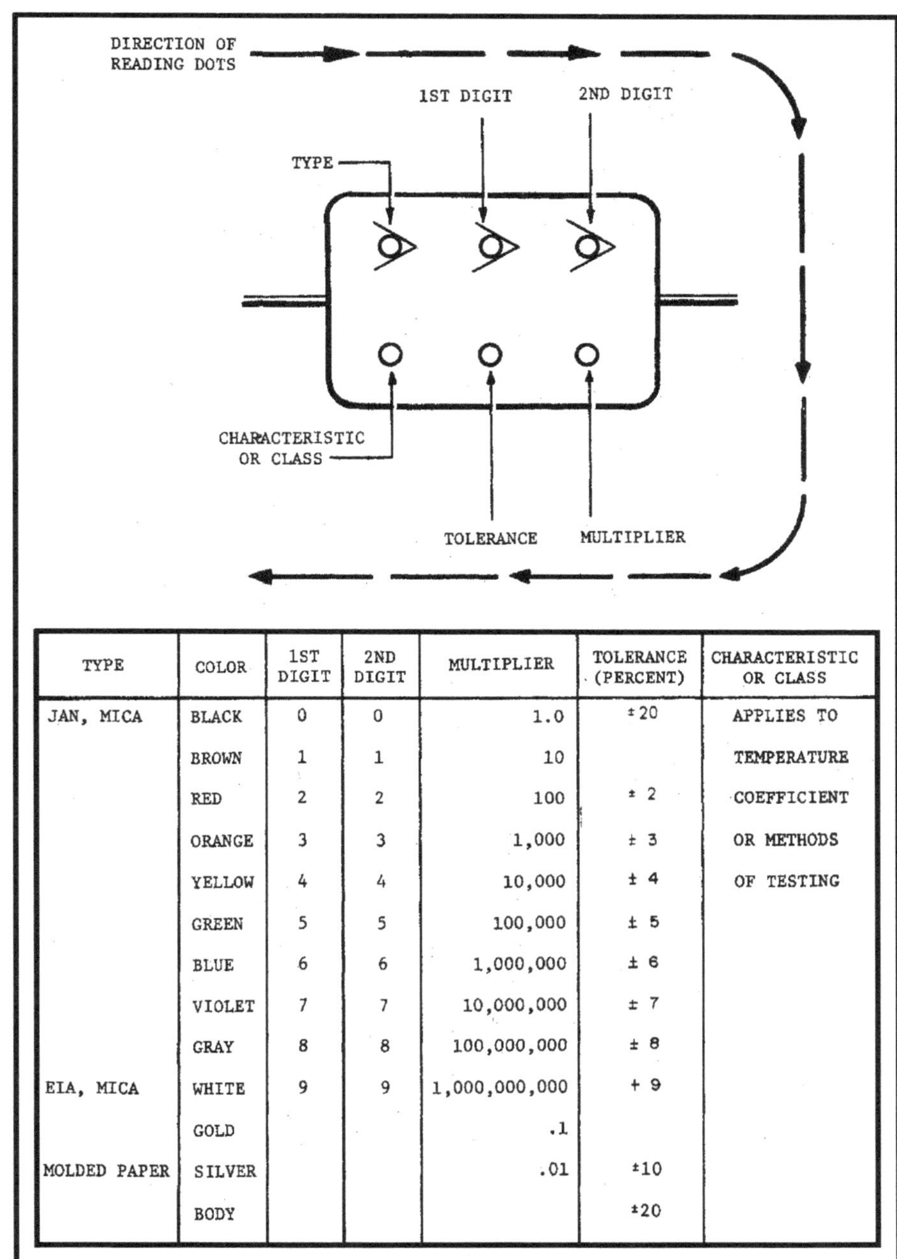

TYPE	COLOR	1ST DIGIT	2ND DIGIT	MULTIPLIER	TOLERANCE (PERCENT)	CHARACTERISTIC OR CLASS
JAN, MICA	BLACK	0	0	1.0	±20	APPLIES TO
	BROWN	1	1	10		TEMPERATURE
	RED	2	2	100	± 2	COEFFICIENT
	ORANGE	3	3	1,000	± 3	OR METHODS
	YELLOW	4	4	10,000	± 4	OF TESTING
	GREEN	5	5	100,000	± 5	
	BLUE	6	6	1,000,000	± 6	
	VIOLET	7	7	10,000,000	± 7	
	GRAY	8	8	100,000,000	± 8	
EIA, MICA	WHITE	9	9	1,000,000,000	+ 9	
	GOLD			.1		
MOLDED PAPER	SILVER			.01	±10	
	BODY				±20	

Figure 3-21.—6-dot color code for mica and molded paper capacitors.

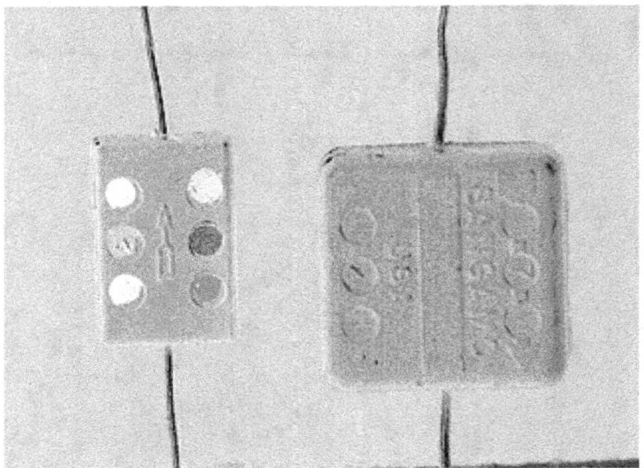

Example of mica capacitors.

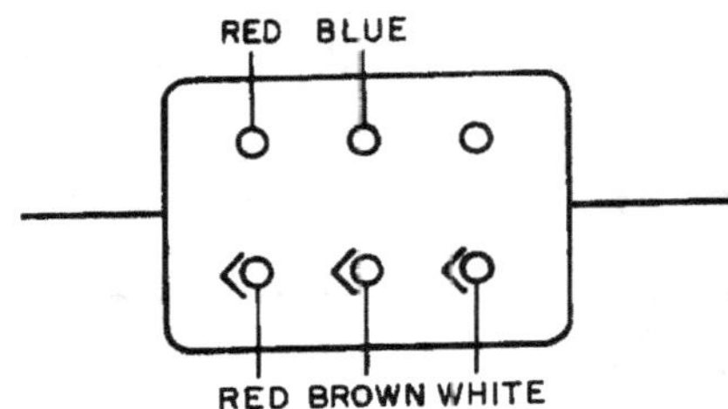

Example of mica capacitors.

To read the capacitor color code on the above capacitor:

1. Hold the capacitor so the arrows point left to right.

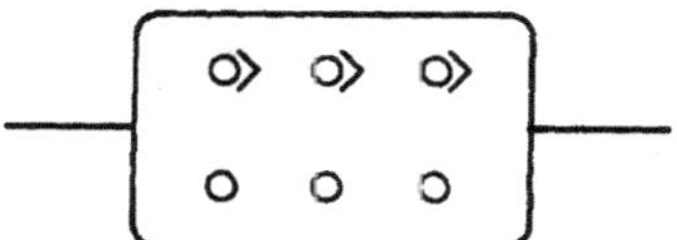

2. Read the first dot.

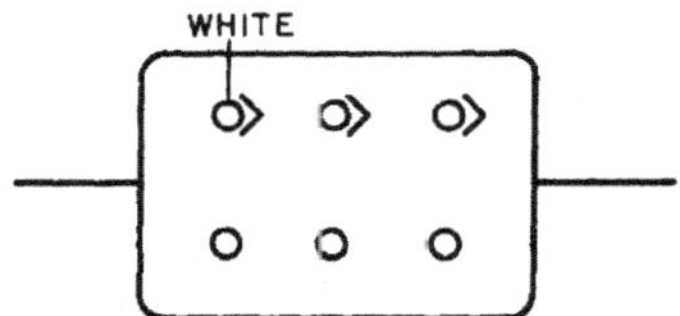

White = mica

3. Read the first digit dot.

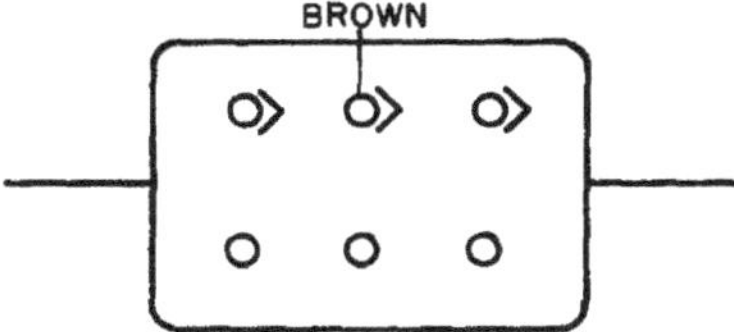

Brown = 1

4. Read the second digit dot and apply it to the first digit.

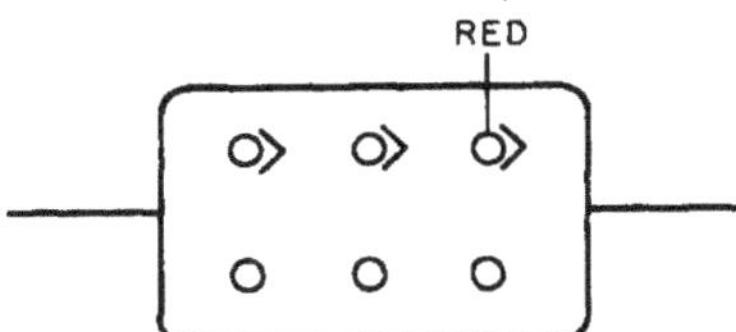

Red = 2 → 12

5. Read the multiplier dot and multiply the first two digits by multiplier. (Remember that the multiplier is in picofarads).

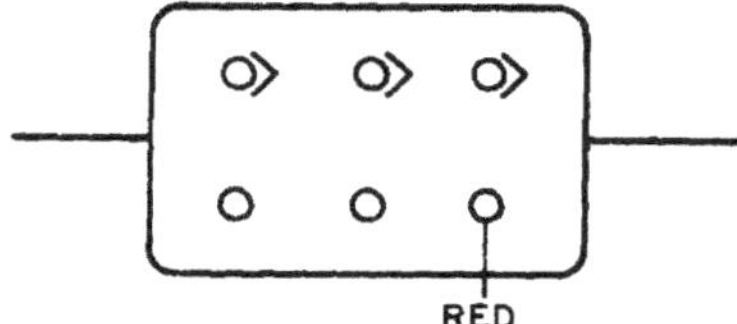

Red = 100 → 12 x 100 = 1200 pF

6. Lastly, read the tolerance dot.

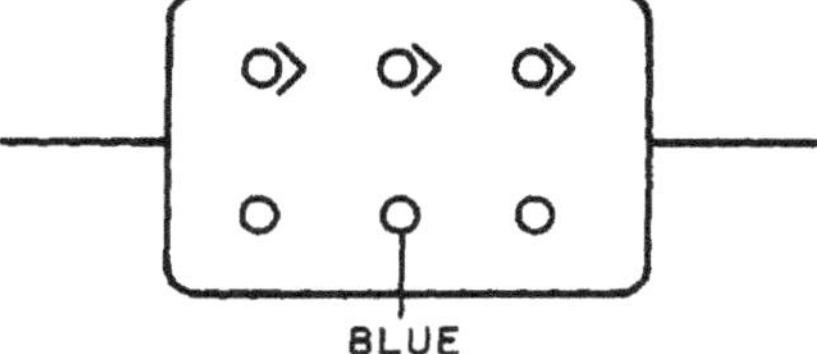

Blue = ±6%

According to the above coding, the capacitor is a mica capacitor whose capacitance is 1200 pF with a tolerance of ±6%.

The capacitor shown in figure 3-22 is a tubular capacitor. Because this type of capacitor always has a paper dielectric, the type code is omitted. To read the code, hold the capacitor so the band closest to the end is on the left side; then read left to right. The last two bands (the fifth and sixth bands from the left) represent the voltage rating of the capacitor. This means that if a capacitor is coded red, red, red, yellow, yellow, yellow, it has the following digit values:

red	=	2
red	=	2
red	=	× 100 pF
yellow	=	±40%
yellow	=	4
yellow	=	4

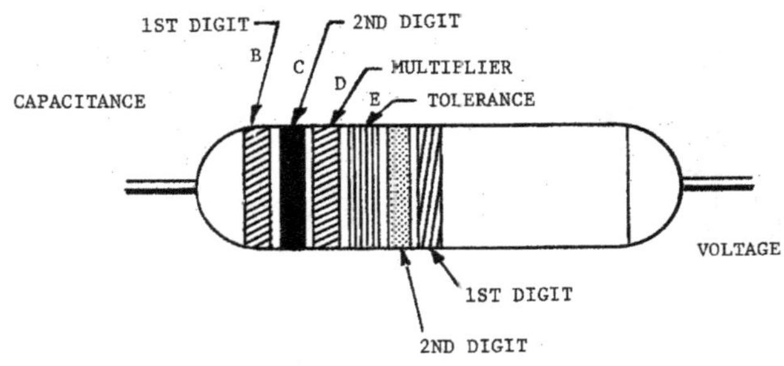

COLOR	CAPACITANCE			TOLERANCE (PERCENT)	VOLTAGE RATING	
	1ST DIGIT	2ND DIGIT	MULTIPLIER		1ST DIGIT	2ND DIGIT
BLACK	0	0	1	±20	0	0
BROWN	1	1	10		1	1
RED	2	2	100		2	2
ORANGE	3	3	1,000	±30	3	3
YELLOW	4	4	10,000	±40	4	4
GREEN	5	5	100,000	± 5	5	5
BLUE	6	6	1,000,000		6	6
VIOLET	7	7			7	7
GRAY	8	8			8	8
WHITE	9	9		±10	9	9

Figure 3-22.—6-band color code for tubular paper dielectric capacitors.

The six digits indicate a capacitance of 2200 pF with a ±40 percent tolerance and a working voltage of 44 volts.

The ceramic capacitor is color coded as shown in figure 3-23 and the mica capacitor as shown in figure 3-24. Notice that this type of mica capacitor differs from the one shown in figure 3-21 in that the arrow is solid instead of broken. This type of mica capacitor is read in the same manner as the one shown in figure 3-21, with one exception: the first dot indicates the first digit. (Note: Because this type of capacitor is always mica, there is no need for a type dot.)

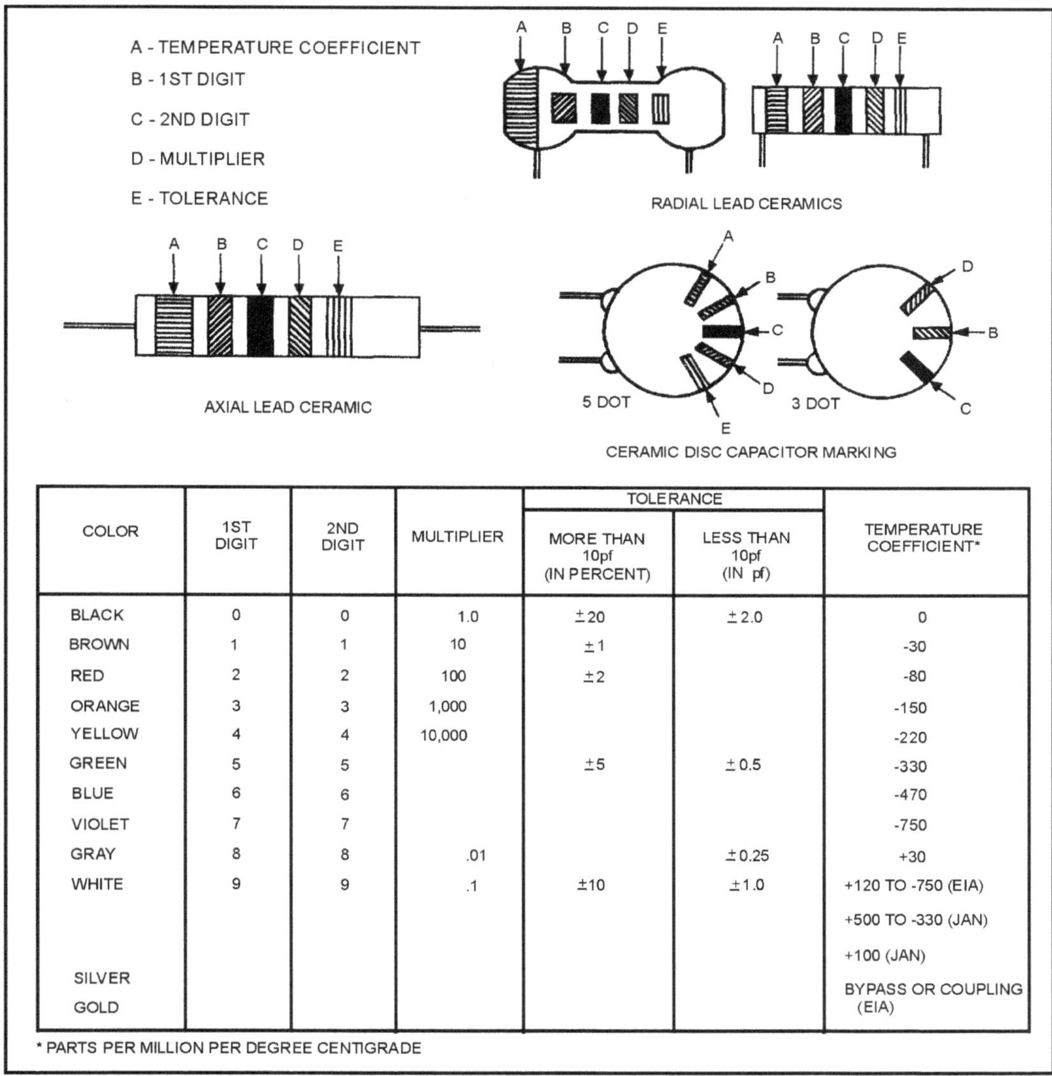

COLOR	1ST DIGIT	2ND DIGIT	MULTIPLIER	TOLERANCE MORE THAN 10pf (IN PERCENT)	TOLERANCE LESS THAN 10pf (IN pf)	TEMPERATURE COEFFICIENT*
BLACK	0	0	1.0	±20	±2.0	0
BROWN	1	1	10	±1		-30
RED	2	2	100	±2		-80
ORANGE	3	3	1,000			-150
YELLOW	4	4	10,000			-220
GREEN	5	5		±5	±0.5	-330
BLUE	6	6				-470
VIOLET	7	7				-750
GRAY	8	8	.01		±0.25	+30
WHITE	9	9	.1	±10	±1.0	+120 TO -750 (EIA)
						+500 TO -330 (JAN)
SILVER						+100 (JAN)
GOLD						BYPASS OR COUPLING (EIA)

* PARTS PER MILLION PER DEGREE CENTIGRADE

Figure 3-23.—Ceramic capacitor color code.

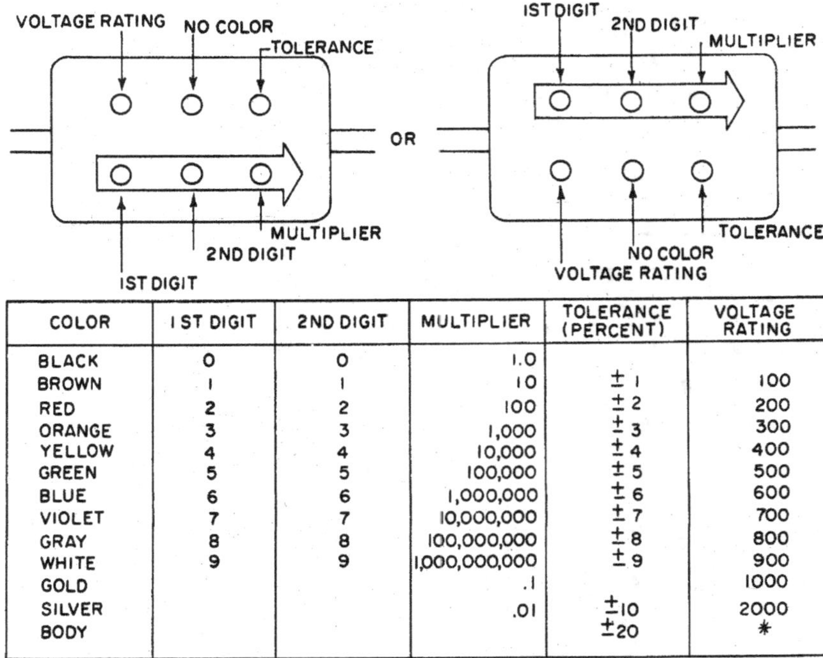

COLOR	I ST DIGIT	2ND DIGIT	MULTIPLIER	TOLERANCE (PERCENT)	VOLTAGE RATING
BLACK	0	0	1.0		
BROWN	1	1	10	± 1	100
RED	2	2	100	± 2	200
ORANGE	3	3	1,000	± 3	300
YELLOW	4	4	10,000	± 4	400
GREEN	5	5	100,000	± 5	500
BLUE	6	6	1,000,000	± 6	600
VIOLET	7	7	10,000,000	± 7	700
GRAY	8	8	100,000,000	± 8	800
WHITE	9	9	1,000,000,000	± 9	900
GOLD			.1		1000
SILVER			.01	±10	2000
BODY				±20	*

* WHERE NO COLOR IS INDICATED.THE VOLTAGE RATING MAY BE AS LOW AS 300 VOLTS.

Figure 3-24.—Mica capacitor color code.

Q19. Examine the three capacitors shown below. What is the capacitance of each?

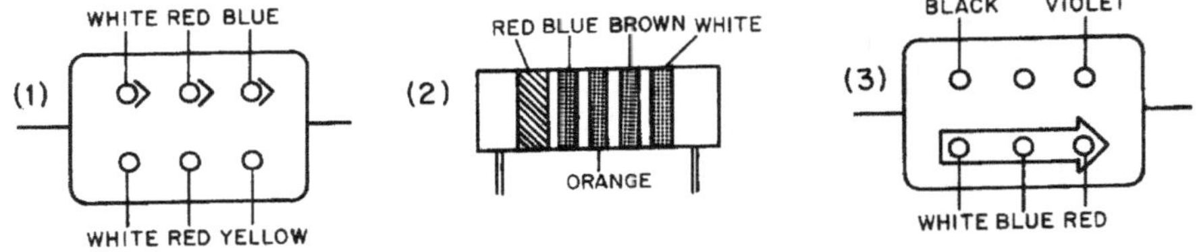

SUMMARY

Before going on to the next chapter, study the below summary to be sure that you understand the important points of this chapter.

THE ELECTROSTATIC FIELD—When a charged body is brought close to another charged body, the bodies either attract or repel one another. (If the charges are alike they repel; if the charges are opposite they attract). The field that causes this effect is called the ELECTROSTATIC FIELD. The amount by which two charges attract or repel each other depends upon the size of the charges and the distance between the charges. The electrostatic field (force between two charged bodies) may be represented by lines of force drawn perpendicular to the charged surfaces. If an electron is placed in the field, it will move toward the positive charge.

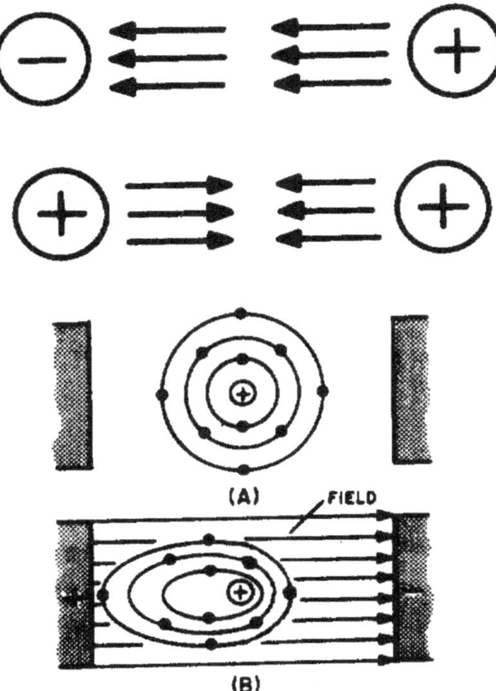

CAPACITANCE—Capacitance is the property of a circuit which OPPOSES any CHANGE in the circuit VOLTAGE. The effect of capacitance may be seen in any circuit where the voltage is changing. Capacitance is usually defined as the ability of a circuit to store electrical energy. This energy is stored in an electrostatic field. The device used in an electrical circuit to store this charge (energy) is called a CAPACITOR. The basic unit of measurement of capacitance is the FARAD (F). A one-farad capacitor will store one coulomb of charge (energy) when a potential of one volt is applied across the capacitor plates. The farad is an enormously large unit of capacitance. More practical units are the microfarad (μF) or the picofarad (pF).

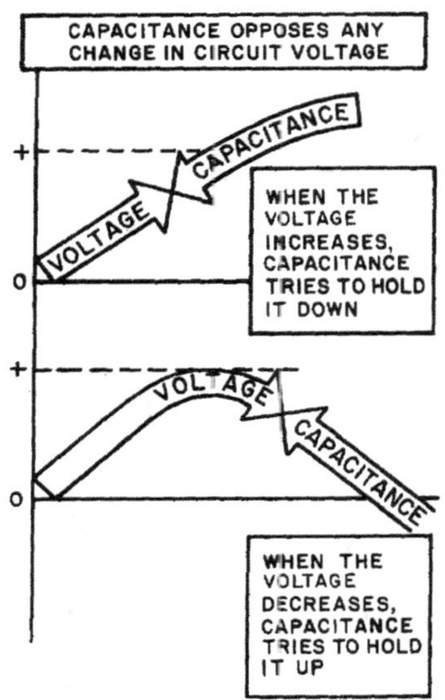

CAPACITOR—A capacitor is a physical device consisting of two pieces of conducting material separated by an insulating material. This insulating material is referred to as the DIELECTRIC. Because the dielectric is an insulator, NO current flows through the capacitor. If the dielectric breaks down and becomes a conductor, the capacitor can no longer hold a charge and is useless. The ability of a dielectric to hold a charge without breaking down is referred to as the dielectric strength. The measure of the ability of the dielectric material to store energy is called the dielectric constant. The dielectric constant is a relative value based on 1.0 for a vacuum.

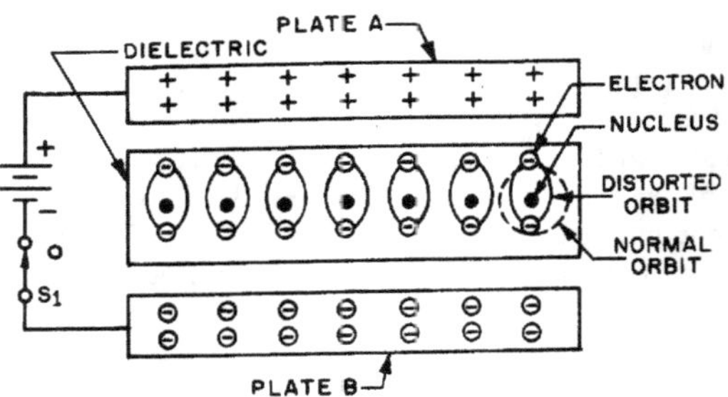

CAPACITORS IN A DC CIRCUIT—When a capacitor is connected to the terminals of a battery, each plate of the capacitor becomes charged. The plate connected to the positive terminal loses electrons. Because this plate has a lack of electrons, it assumes a positive charge. The plate connected to the negative terminal gains electrons. Because the plate has an excess of electrons, it assumes a negative charge. This process continues until the charge across the plates equals the applied voltage. At this point current ceases to flow in the circuit. As long as nothing changes in the circuit, the capacitor will hold its charge and there will be no current in any part of the circuit. If the leads of the capacitor are now shorted together, current again flows in the circuit. Current will continue to flow until the charges on the two plates become equal. At this point, current ceases to flow. With a dc voltage source, current will flow in the circuit only long enough to charge (or discharge) the capacitor. Thus, a capacitor does NOT allow dc current to flow continuously in a circuit.

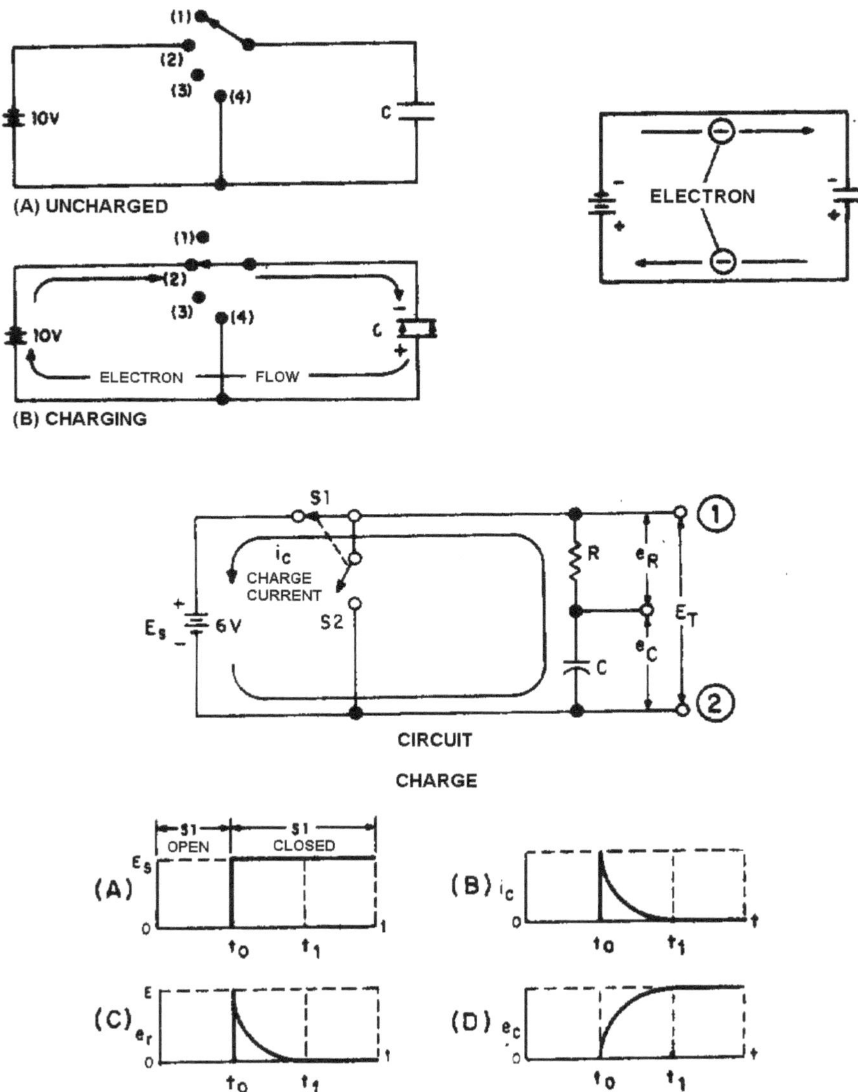

FACTORS AFFECTING CAPACITANCE—There are three factors affecting capacitance. One factor is the area of the plate surfaces. Increasing the area of the plate increases the capacitance. Another factor is the amount of space between the plates. The closer the plates, the greater will be the electrostatic field. A greater electrostatic field causes a greater capacitance. The plate spacing is determined by the thickness of the dielectric. The third factor affecting capacitance is the dielectric constant. The value of the dielectric constant is dependent upon the type of dielectric used.

WORKING VOLTAGE—The working voltage of a capacitor is the maximum voltage that can be steadily applied to the capacitor without the capacitor breaking down (shorting). The working voltage depends upon the type of material used as the dielectric (the dielectric constant) and the thickness of the dielectric.

CAPACITOR LOSSES—Power losses in a capacitor are caused by dielectric leakage and dielectric hysteresis. Dielectric leakage loss is caused by the leakage current through the resistance in the dielectric. Although this resistance is extremely high, a small amount of current does flow. Dielectric hysteresis may be defined as an effect in a dielectric material similar to the hysteresis found in a magnetic material.

RC TIME CONSTANT—The time required to charge a capacitor to 63.2 percent of the applied voltage, or to discharge the capacitor to 36.8 percent of its charge. The time constant (t) is equal to the product of the resistance and the capacitance. Expressed as a formula:

$$t = RC$$

where t is in seconds, R is in ohms, and C is in farads.

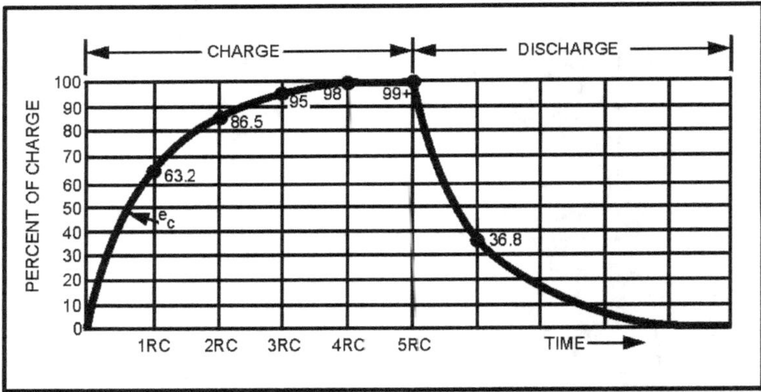

CAPACITORS IN SERIES—The effect of wiring capacitors in series is to increase the distance between plates. This reduces the total capacitance of the circuit. Total capacitance for series connected capacitors may be computed by the formula:

$$C_T = \frac{1}{\frac{1}{C1} + \frac{1}{C2} + \frac{1}{C3} + \ldots \frac{1}{C_n}}$$

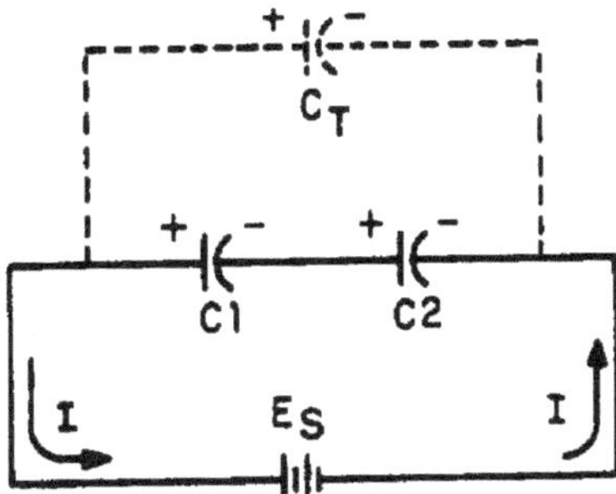

If an electrical circuit contains only two series connected capacitors, C_T may be computed using the following formula:

$$C_T = \frac{C1\,C2}{C1 + C2}$$

CAPACITORS IN PARALLEL—The effect of wiring capacitors in parallel is to increase the plate area of the capacitors. Total capacitance (C_T) may be found using the formula:

$$C_T = C1 + C2 \ldots + C_n$$

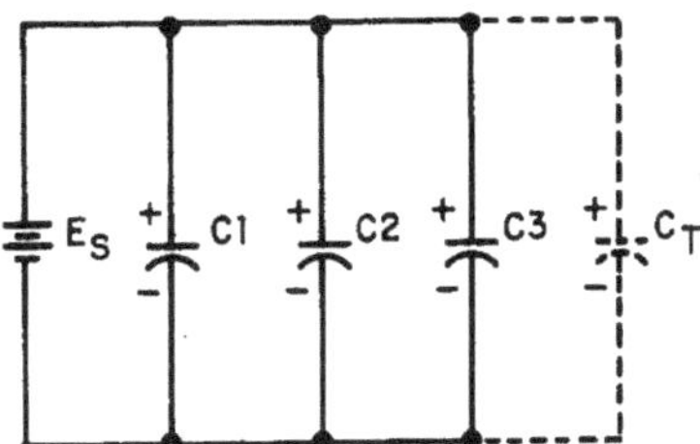

TYPES OF CAPACITORS—Capacitors are manufactured in various forms and may be divided into two main classes-fixed capacitors and variable capacitors. A fixed capacitor is constructed to have a constant or fixed value of capacitance. A variable capacitor allows the capacitance to be varied or adjusted.

ANSWERS TO QUESTIONS Q1. THROUGH Q19.

A1.

a. A capacitor is a device that stores electrical energy in an electrostatic field.

b. Capacitance is the property of a circuit which opposes changes in voltage.

A2.

a. They are polarized from positive to negative.

b. They radiate from a charged particle in straight lines and do not form closed loops.

c. They have the ability to pass through any known material.

d. They have the ability to distort the orbits of electrons circling the nucleus.

A3. Toward the positive charge.

A4. Two pieces of conducting material separated by an insulator.

A5. A farad is the unit of capacitance. A capacitor has a capacitance of 1 farad when a difference of 1 volt will charge it with 1 coulomb of electrons.

A6.

a. One microfarad equals 10^{-6} farad.

b. One picofarad equals 10^{-12} farad.

A7.

a. The area of the plates.

b. The distance between the plates.

c. The dielectric constant of the material between the plates.

A8.

4372 picofarads

$$C = .2249\left(\frac{KA}{d}\right)$$

$$C = .2249\left(\frac{81 \times 6}{.025}\right)$$

$$C = 4372 \text{ (Rounded off)}$$

A9.

a. Hysteresis

b. Dielectric leakage

A10.

a. It is the maximum voltage the capacitor can work without risk of damage.

b. 900 volts.

A11.

a. When the capacitor is charging, electrons accumulate on the negative plate and leave the positive plate until the charge on the capacitor is equal to the battery voltage.

b. When the capacitor is discharging, electrons flow from the negatively charged plate to the positively charged plate until the charge on each plate is neutral.

A12. *At the instant of the initiation of the action.*

A13. *Zero.*

A14.

144 seconds

$$t = R\,(\text{megohms}) \times C\,(\text{microfarads})$$

$$t = 12 \times 12$$

$$t = 144\ \text{seconds}$$

A15.

.01 microfarads 40% from the graph = .5

$$RC = \frac{200}{.5}$$

$$RC = 400\ \text{microseconds}$$

$$C = \frac{t}{R}$$

$$C = \frac{400\,\mu s}{40{,}000\,\Omega}$$

$$C = .01\,\mu F = 10{,}000\,pF$$

A16.

$.1\mu F$

$$C_T = \frac{C_1 C_2}{C_1 + C_2}$$

$$C_T = \frac{10 \times 0.1}{10 + 0.1}\mu F$$

$$C_T = \frac{1}{10.1}\mu F$$

$$C_T = .099\,\mu F \text{ or } 0.1\mu F$$

A17.

$33.1\mu F$

$$C_T = C1 + C2 + C3 + C4$$

$$C_T = 10\,\mu F + 21\,\mu F + 0.1\mu F + 2\mu F$$

$$C_T = 33.1\mu F$$

A18.

a. *Electrolytic capacitor*

b. *Trimmer capacitor*

A19.

a. *26 μF or 260,000 pF*

b. *630 pF*

c. *9600 pF*

CHAPTER 4

INDUCTIVE AND CAPACITIVE REACTANCE

LEARNING OBJECTIVES

Upon completion of this chapter you will be able to:

1. State the effects an inductor has on a change in current and a capacitor has on a change in voltage.
2. State the phase relationships between current and voltage in an inductor and in a capacitor.
3. State the terms for the opposition an inductor and a capacitor offer to ac
4. Write the formulas for inductive and capacitive reactances.
5. State the effects of a change in frequency on X_L and X_C.
6. State the effects of a change in inductance on X_L and a change in capacitance on X_C.
7. Write the formula for determining total reactance (X); compute total reactance (X) in a series circuit; and indicate whether the total reactance is capacitive or inductive.
8. State the term given to the total opposition (Z) in an ac circuit.
9. Write the formula for impedance, and calculate the impedance in a series circuit when the values of X_C, X_L, and R are given.
10. Write the Ohm's law formulas used to determine voltage and current in an ac circuit.
11. Define true power, reactive power, and apparent power; state the unit of measurement for and the formula used to calculate each.
12. State the definition of and write the formula for power factor.
13. Given the power factor and values of X and R in an ac circuit, compute the value of reactance in the circuit, and state the type of reactance that must be connected in the circuit to correct the power factor to unity (1).
14. State the difference between calculating impedance in a series ac circuit and in a parallel ac circuit.

INDUCTIVE AND CAPACITIVE REACTANCE

You have already learned how inductance and capacitance individually behave in a direct current circuit. In this chapter you will be shown how inductance, capacitance, and resistance affect alternating current.

INDUCTANCE AND ALTERNATING CURRENT

This might be a good place to recall what you learned about phase in chapter 1. When two things are in step, going through a cycle together, falling together and rising together, they are in phase. When they are out of phase, the angle of lead or lag-the number of electrical degrees by which one of the values leads or lags the other-is a measure of the amount they are out of step. The time it takes the current in an inductor to build up to maximum and to fall to zero is important for another reason. It helps illustrate a very useful characteristic of inductive circuits-the current through the inductor always lags the voltage across the inductor.

A circuit having pure resistance (if such a thing were possible) would have the alternating current through it and the voltage across it rising and failing together. This is illustrated in figure 4-1(A), which shows the sine waves for current and voltage in a purely resistive circuit having an ac source. The current and voltage do not have the same amplitude, but they are in phase.

In the case of a circuit having inductance, the opposing force of the counter emf would be enough to keep the current from remaining in phase with the applied voltage. You learned that in a dc circuit containing pure inductance the current took time to rise to maximum even though the full applied voltage was immediately at maximum. Figure 4-1(B) shows the wave forms for a purely inductive ac circuit in steps of quarter-cycles.

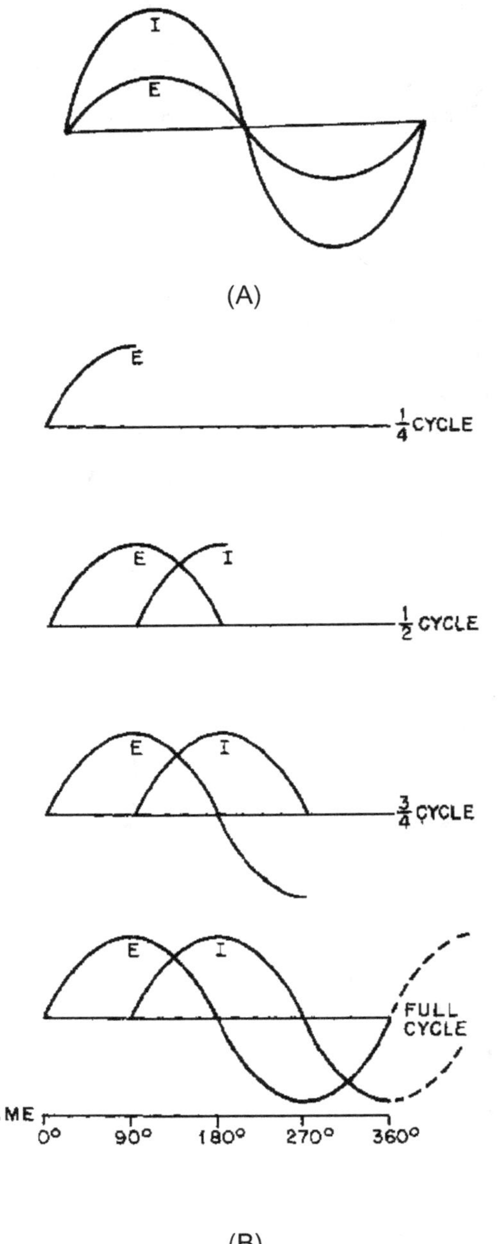

Figure 4-1.—Voltage and current waveforms in an inductive circuit.

With an ac voltage, in the first quarter-cycle (0° to 90°) the applied ac voltage is continually increasing. If there was no inductance in the circuit, the current would also increase during this first quarter-cycle. You know this circuit does have inductance. Since inductance opposes any change in current flow, no current flows during the first quarter-cycle. In the next quarter-cycle (90° to 180°) the voltage decreases back to zero; current begins to flow in the circuit and reaches a maximum value at the same instant the voltage reaches zero. The applied voltage now begins to build up to maximum in the other direction, to be followed by the resulting current. When the voltage again reaches its maximum at the end of the third quarter-cycle (270°) all values are exactly opposite to what they were during the first half-cycle. The applied voltage leads the resulting current by one quarter-cycle or 90 degrees. To complete the full 360° cycle of the voltage, the voltage again decreases to zero and the current builds to a maximum value.

You must not get the idea that any of these values stops cold at a particular instant. Until the applied voltage is removed, both current and voltage are always changing in amplitude and direction.

As you know the sine wave can be compared to a circle. Just as you mark off a circle into 360 degrees, you can mark off the time of one cycle of a sine wave into 360 electrical degrees. This relationship is shown in figure 4-2. By referring to this figure you can see why the current is said to lag the voltage, in a purely inductive circuit, by 90 degrees. Furthermore, by referring to figures 4-2 and 4-1(A) you can see why the current and voltage are said to be in phase in a purely resistive circuit. In a circuit having both resistance and inductance then, as you would expect, the current lags the voltage by an amount somewhere between 0 and 90 degrees.

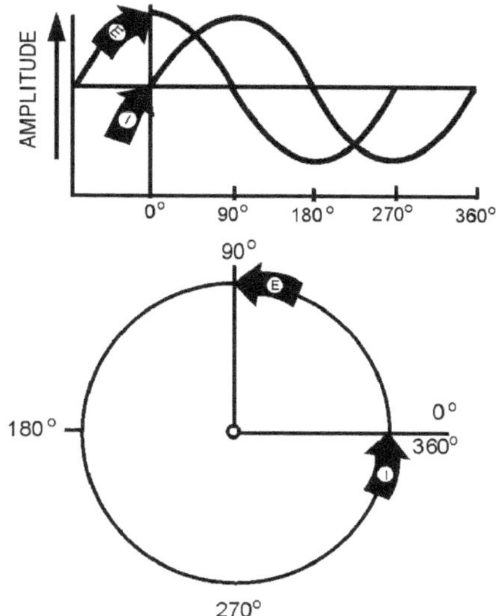

Figure 4-2.—Comparison of sine wave and circle in an inductive circuit.

A simple memory aid to help you remember the relationship of voltage and current in an inductive circuit is the word ELI. Since E is the symbol for voltage, L is the symbol for inductance, and I is the symbol for current; the word ELI demonstrates that current comes after (Lags) voltage in an inductor.

Q1. What effect does an inductor have on a change in current?

Q2. What is the phase relationship between current and voltage in an inductor?

INDUCTIVE REACTANCE

When the current flowing through an inductor continuously reverses itself, as in the case of an ac source, the inertia effect of the cemf is greater than with dc. The greater the amount of inductance (L), the greater the opposition from this inertia effect. Also, the faster the reversal of current, the greater this inertial opposition. This opposing force which an inductor presents to the FLOW of alternating current cannot be called resistance, since it is not the result of friction within a conductor. The name given to it is INDUCTIVE REACTANCE because it is the "reaction" of the inductor to the changing value of alternating current. Inductive reactance is measured in ohms and its symbol is X_L.

As you know, the induced voltage in a conductor is proportional to the rate at which magnetic lines of force cut the conductor. The greater the rate (the higher the frequency), the greater the cemf. Also, the induced voltage increases with an increase in inductance; the more ampere-turns, the greater the cemf. Reactance, then, increases with an increase of frequency and with an increase of inductance. The formula for inductive reactance is as follows:

$$X_L = 2\pi fL$$

Where:

X_L is inductive reactance in ohms.

2π is a constant in which the Greek letter π, called "pi" represents 3.1416 and $2 \times \pi =$ 6.28 approximately.

f is frequency of the alternating current in Hz.

L is inductance in henrys.

The following example problem illustrates the computation of X_L.

Given: $f = 60$ Hz

$L = 20$ H

Solution: $X_L = 2\pi fL$

$X_L = 6.28 \times 60 \text{ Hz} \times 20 \text{ H}$

$X_L = 7{,}536\ \Omega$

Q3. What is the term for the opposition an inductor presents to ac?

Q4. What is the formula used to compute the value of this opposition?

Q5. What happens to the value of X_L as frequency increases?

Q6. What happens to the value of X_L as inductance decreases?

CAPACITORS AND ALTERNATING CURRENT

The four parts of figure 4-3 show the variation of the alternating voltage and current in a capacitive circuit, for each quarter of one cycle. The solid line represents the voltage across the capacitor, and the dotted line represents the current. The line running through the center is the zero, or reference point, for both the voltage and the current. The bottom line marks off the time of the cycle in terms of electrical degrees. Assume that the ac voltage has been acting on the capacitor for some time before the time represented by the starting point of the sine wave in the figure.

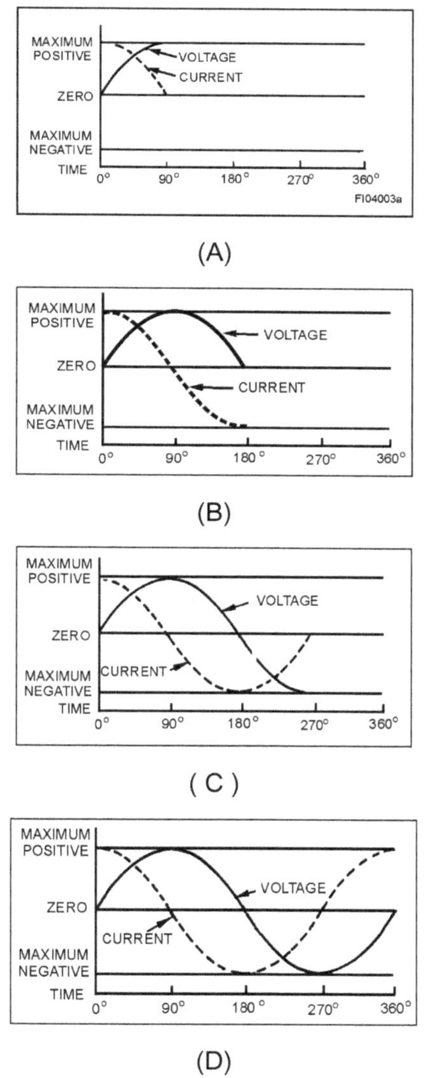

Figure 4-3.—Phase relationship of voltage and current in a capacitive circuit.

At the beginning of the first quarter-cycle (0° to 90°) the voltage has just passed through zero and is increasing in the positive direction. Since the zero point is the steepest part of the sine wave, the voltage is changing at its greatest rate. The charge on a capacitor varies directly with the voltage, and therefore the charge on the capacitor is also changing at its greatest rate at the beginning of the first quarter-cycle. In other words, the greatest number of electrons are moving off one plate and onto the other plate. Thus the capacitor current is at its maximum value, as part (A) of the figure shows.

As the voltage proceeds toward maximum at 90 degrees, its rate of change becomes less and less, hence the current must decrease toward zero. At 90 degrees the voltage across the capacitor is maximum, the capacitor is fully charged, and there is no further movement of electrons from plate to plate. That is why the current at 90 degrees is zero.

At the end of this first quarter-cycle the alternating voltage stops increasing in the positive direction and starts to decrease. It is still a positive voltage, but to the capacitor the decrease in voltage means that the plate which has just accumulated an excess of electrons must lose some electrons. The current flow, therefore, must reverse its direction. Part (B) of the figure shows the current curve to be below the zero line (negative current direction) during the second quarter-cycle (90° to 180°).

At 180 degrees the voltage has dropped to zero. This means that for a brief instant the electrons are equally distributed between the two plates; the current is maximum because the rate of change of voltage is maximum. Just after 180 degrees the voltage has reversed polarity and starts building up its maximum negative peak which is reached at the end of the third quarter-cycle (180° to 270°). During this third quarter-cycle the rate of voltage change gradually decreases as the charge builds to a maximum at 270 degrees. At this point the capacitor is fully charged and it carries the full impressed voltage. Because the capacitor is fully charged there is no further exchange of electrons; therefore, the current flow is zero at this point. The conditions are exactly the same as at the end of the first quarter-cycle (90°) but the polarity is reversed.

Just after 270 degrees the impressed voltage once again starts to decrease, and the capacitor must lose electrons from the negative plate. It must discharge, starting at a minimum rate of flow and rising to a maximum. This discharging action continues through the last quarter-cycle (270° to 360°) until the impressed-voltage has reached zero. At 360 degrees you are back at the beginning of the entire cycle, and everything starts over again.

If you examine the complete voltage and current curves in part D, you will see that the current always arrives at a certain point in the cycle 90 degrees ahead of the voltage, because of the charging and discharging action. You know that this time and place relationship between the current and voltage is called the phase relationship. The voltage-current phase relationship in a capacitive circuit is exactly opposite to that in an inductive circuit. The current of a capacitor leads the voltage across the capacitor by 90 degrees.

You realize that the current and voltage are both going through their individual cycles at the same time during the period the ac voltage is impressed. The current does not go through part of its cycle (charging or discharging), stop, and wait for the voltage to catch up. The amplitude and polarity of the voltage and the amplitude and direction of the current are continually changing. Their positions with respect to each other and to the zero line at any electrical instant-any degree between zero and 360 degrees-can be seen by reading upwards from the time-degree line. The current swing from the positive peak at zero degrees to the negative peak at 180 degrees is NOT a measure of the number of electrons, or the charge on the plates. It is a picture of the direction and strength of the current in relation to the polarity and strength of the voltage appearing across the plates.

At times it is convenient to use the word "ICE" to recall to mind the phase relationship of the current and voltage in capacitive circuits. I is the symbol for current, and in the word ICE it leads, or comes before, the symbol for voltage, E. C, of course, stands for capacitor. This memory aid is similar to the "ELI" used to remember the current and voltage relationship in an inductor. The phrase "ELI the ICE man" is helpful in remembering the phase relationship in both the inductor and capacitor.

Since the plates of the capacitor are changing polarity at the same rate as the ac voltage, the capacitor seems to pass an alternating current. Actually, the electrons do not pass through the dielectric, but their rushing back and forth from plate to plate causes a current flow in the circuit. It is convenient, however, to say that the alternating current flows "through" the capacitor. You know this is not true, but the expression avoids a lot of trouble when speaking of current flow in a circuit containing a capacitor. By the same short cut, you may say that the capacitor does not pass a direct current (if both plates are connected to a dc source, current will flow only long enough to charge the capacitor). With a capacitor type of hookup in a circuit containing both ac and dc, only the ac will be "passed" on to another circuit.

You have now learned two things to remember about a capacitor: A capacitor will appear to conduct an alternating current and a capacitor will not conduct a direct current.

Q7. What effect does the capacitor have on a changing voltage?

Q8. What is the phase relationship between current and voltage in a capacitor?

CAPACITIVE REACTANCE

So far you have been dealing with the capacitor as a device which passes ac and in which the only opposition to the alternating current has been the normal circuit resistance present in any conductor. However, capacitors themselves offer a very real opposition to current flow. This opposition arises from the fact that, at a given voltage and frequency, the number of electrons which go back and forth from plate to plate is limited by the storage ability-that is, the capacitance-of the capacitor. As the capacitance is increased, a greater number of electrons change plates every cycle, and (since current is a measure of the number of electrons passing a given point in a given time) the current is increased.

Increasing the frequency will also decrease the opposition offered by a capacitor. This occurs because the number of electrons which the capacitor is capable of handling at a given voltage will change plates more often. As a result, more electrons will pass a given point in a given time (greater current flow). The opposition which a capacitor offers to ac is therefore inversely proportional to frequency and to capacitance. This opposition is called CAPACITIVE REACTANCE. You may say that capacitive reactance decreases with increasing frequency or, for a given frequency, the capacitive reactance decreases with increasing capacitance. The symbol for capacitive reactance is X_C.

Now you can understand why it is said that the X_C varies inversely with the product of the frequency and capacitance. The formula is:

$$X_C = \frac{1}{2\pi fC}$$

Where:

X_C is capacitive reactance in ohms

f is frequency in Hertz

C is capacitance in farads

π is 6.28 (2 × 3.1416)

The following example problem illustrates the computation of X_C.

Given: f = 100 Hz
C = 50 μF

Solution:

$$X_C = \frac{1}{2\pi fC}$$

$$X_C = \frac{1}{6.28 \times 100\ \text{Hz} \times 50\ \mu\text{F}}$$

$$X_C = \frac{1}{.0314}\Omega$$

$$X_C = 31.8\,\Omega \text{ or } 32\,\Omega$$

Q9. What is the term for the opposition that a capacitor presents to ac?

Q10. What is the formula used to compute this opposition?

Q11. What happens to the value of X_C as frequency decreases?

Q12. What happens to the value of X_C as capacitance increases?

REACTANCE, IMPEDANCE, AND POWER RELATIONSHIPS IN AC CIRCUITS

Up to this point inductance and capacitance have been explained individually in ac circuits. The rest of this chapter will concern the combination of inductance, capacitance, and resistance in ac circuits.

To explain the various properties that exist within ac circuits, the series RLC circuit will be used. Figure 4-4 is the schematic diagram of the series RLC circuit. The symbol shown in figure 4-4 that is marked E is the general symbol used to indicate an ac voltage source.

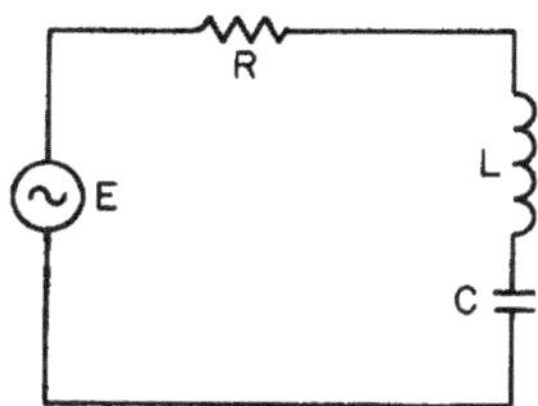

Figure 4-4.—Series RLC circuit.

REACTANCE

The effect of inductive reactance is to cause the current to lag the voltage, while that of capacitive reactance is to cause the current to lead the voltage. Therefore, since inductive reactance and capacitive reactance are exactly opposite in their effects, what will be the result when the two are combined? It is not hard to see that the net effect is a tendency to cancel each other, with the combined effect then equal to the difference between their values. This resultant is called REACTANCE; it is represented by the symbol X; and expressed by the equation $X = X_L - X_C$ or $X = X_C - X_L$. Thus, if a circuit contains 50 ohms of inductive reactance and 25 ohms of capacitive reactance in series, the net reactance, or X, is 50 ohms – 25 ohms, or 25 ohms of inductive reactance.

For a practical example, suppose you have a circuit containing an inductor of 100 μH in series with a capacitor of .001 μF, and operating at a frequency of 4 MHz. What is the value of net reactance, or X?

Given:

$$f = 4\,MHz$$
$$L = 100\,\mu H$$
$$C = .001\,\mu F$$

Solution:

$$X_L = 2\pi fL$$
$$X_L = 6.28 \times 4\,MHz \times 100\,\mu H$$
$$X_L = 2512\,\Omega$$
$$X_C = \frac{1}{2\pi fC}$$
$$X_C = \frac{1}{6.28 \times 4\,MHz \times .001\,\mu F}$$
$$X_C = \frac{1}{.02512}\,\Omega$$
$$X_C = 39.8\,\Omega$$
$$X = X_L - X_C$$
$$X = 2512\,\Omega - 39.8\,\Omega$$
$$X = 2472.2\,\Omega \text{ (inductive)}$$

Now assume you have a circuit containing a 100 - μH inductor in series with a .0002-μF capacitor, and operating at a frequency of 1 MHz. What is the value of the resultant reactance in this case?

Given: $f = 1\,MHz$

$L = 100\,\mu H$

$C = .0002\,\mu F$

Solution:

$$X_L = 2\pi fL$$

$$X_L = 6.28 \times 1\,MHz \times 100\,\mu H$$

$$X_L = 628\,\Omega$$

$$X_C = \frac{1}{2\pi fC}$$

$$X_C = \frac{1}{6.28 \times 1\,MHz \times .0002\,\mu F}$$

$$X_C = \frac{1}{.001256}\,\Omega$$

$$X_C = 796\,\Omega$$

$$X = X_C - X_L$$

$$X = 796\,\Omega - 628\,\Omega$$

$$X = 168\,\Omega \text{ (capacitive)}$$

You will notice that in this case the inductive reactance is smaller than the capacitive reactance and is therefore subtracted from the capacitive reactance.

These two examples serve to illustrate an important point: when capacitive and inductive reactance are combined in series, the smaller is always subtracted from the larger and the resultant reactance always takes the characteristics of the larger.

Q13. What is the formula for determining total reactance in a series circuit where the values of X_C and X_L are known?

Q14. What is the total amount of reactance (X) in a series circuit which contains an X_L of 20 ohms and an X_C of 50 ohms? (Indicate whether X is capacitive or inductive)

IMPEDANCE

From your study of inductance and capacitance you know how inductive reactance and capacitive reactance act to oppose the flow of current in an ac circuit. However, there is another factor, the resistance, which also opposes the flow of the current. Since in practice ac circuits containing reactance also contain resistance, the two combine to oppose the flow of current. This combined opposition by the resistance and the reactance is called the IMPEDANCE, and is represented by the symbol Z.

Since the values of resistance and reactance are both given in ohms, it might at first seem possible to determine the value of the impedance by simply adding them together. It cannot be done so easily, however. You know that in an ac circuit which contains only resistance, the current and the voltage will be in step (that is, in phase), and will reach their maximum values at the same instant. You also know that in an ac circuit containing only reactance the current will either lead or lag the voltage by one-quarter of a cycle or 90 degrees. Therefore, the voltage in a purely reactive circuit will differ in phase by 90 degrees from that in a purely resistive circuit and for this reason reactance and resistance are rot combined by simply adding them.

When reactance and resistance are combined, the value of the impedance will be greater than either. It is also true that the current will not be in step with the voltage nor will it differ in phase by exactly 90 degrees from the voltage, but it will be somewhere between the in-step and the 90-degree out-of-step conditions. The larger the reactance compared with the resistance, the more nearly the phase difference will approach 90°. The larger the resistance compared to the reactance, the more nearly the phase difference will approach zero degrees.

If the value of resistance and reactance cannot simply be added together to find the impedance, or Z, how is it determined? Because the current through a resistor is in step with the voltage across it and the current in a reactance differs by 90 degrees from the voltage across it, the two are at right angles to each other. They can therefore be combined by means of the same method used in the construction of a right-angle triangle.

Assume you want to find the impedance of a series combination of 8 ohms resistance and 5 ohms inductive reactance. Start by drawing a horizontal line, R, representing 8 ohms resistance, as the base of the triangle. Then, since the effect of the reactance is always at right angles, or 90 degrees, to that of the resistance, draw the line X_L, representing 5 ohms inductive reactance, as the altitude of the triangle. This is shown in figure 4-5. Now, complete the hypotenuse (longest side) of the triangle. Then, the hypotenuse represents the impedance of the circuit.

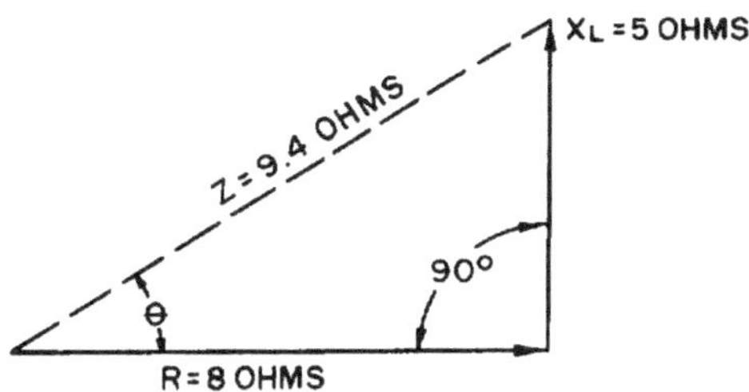

Figure 4-5.—Vector diagram showing relationship of resistance, inductive reactance, and impedance in a series circuit.

One of the properties of a right triangle is:

$$(\text{hypotenuse})^2 = (\text{base})^2 + (\text{altitude})^2$$

or,

$$\text{hypotenuse} = \sqrt{(\text{base})^2 + (\text{altitude})^2}$$

Applied to impedance, this becomes,

$$(\text{impedance})^2 = (\text{resistance})^2 + (\text{reactance})^2$$

or,

$$\text{impedance} = \sqrt{(\text{resistance})^2 + (\text{reactance})^2}$$

or,

$$Z = \sqrt{R^2 + X^2}$$

Now suppose you apply this equation to check your results in the example given above.

Given: $R = 8\,\Omega$

$X_L = 5\,\Omega$

Solution:

$$Z = \sqrt{R^2 + X_L^2}$$

$$Z = \sqrt{(8\,\Omega)^2 + (5\,\Omega)^2}$$

$$Z = \sqrt{64 + 25\,\Omega}$$

$$Z = \sqrt{89\,\Omega}$$

$$Z = 9.4\,\Omega$$

(See the Appendix III for a square Root Table.)

When you have a capacitive reactance to deal with instead of inductive reactance as in the previous example, it is customary to draw the line representing the capacitive reactance in a downward direction. This is shown in figure 4-6. The line is drawn downward for capacitive reactance to indicate that it acts in a direction opposite to inductive reactance which is drawn upward. In a series circuit containing capacitive reactance the equation for finding the impedance becomes:

$$Z = \sqrt{R^2 + X_C^2}$$

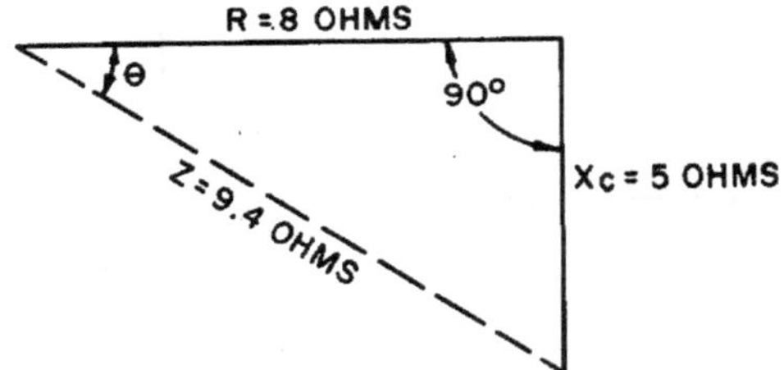

Figure 4-6.—Vector diagram showing relationship of resistance, capacitive reactance, and impedance in a series circuit.

In many series circuits you will find resistance combined with both inductive reactance and capacitive reactance. Since you know that the value of the reactance, X, is equal to the difference between the values of the inductive reactance, X_L, and the capacitive reactance, X_C, the equation for the impedance in a series circuit containing R, X_L, and X_C then becomes:

$$Z = \sqrt{R^2 + (X_L - X_C)^2}$$

or,

$$Z = \sqrt{R^2 + X^2}$$

(Note: The formulas $Z = \sqrt{R^2 + X_L^2}$, $Z = \sqrt{R^2 + X_C^2}$, and $Z = \sqrt{R^2 + X^2}$ can be used to calculate Z only if the resistance and reactance are connected in series.)

In figure 4-7 you will see the method which may be used to determine the impedance in a series circuit consisting of resistance, inductance, and capacitance.

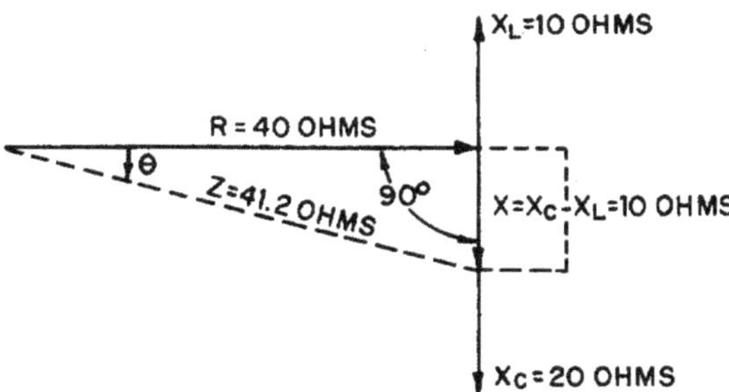

Figure 4-7.—Vector diagram showing relationship of resistance, reactance (capacitive and inductive), and impedance in a series circuit.

Assume that 10 ohms inductive reactance and 20 ohms capacitive reactance are connected in series with 40 ohms resistance. Let the horizontal line represent the resistance R. The line drawn upward from the end of R, represents the inductive reactance, X_L. Represent the capacitive reactance by a line drawn downward at right angles from the same end of R. The resultant of X_L and X_C is found by subtracting X_L from X_C. This resultant represents the value of X.

Thus:

$$X = X_C - X_L$$
$$X = 10 \text{ ohms}$$

The line, Z, will then represent the resultant of R and X. The value of Z can be calculated as follows:

Given: $X_L = 10\ \Omega$
$X_C = 20\ \Omega$
$R = 40\ \Omega$

Solution:

$$X = X_c - X_L$$
$$X = 20\,\Omega - 10\,\Omega$$
$$X = 10\,\Omega$$
$$Z = \sqrt{R^2 + X^2}$$
$$Z = \sqrt{(40\,\Omega)^2 + (10\,\Omega^2)}$$
$$Z = \sqrt{1600 + 100\,\Omega}$$
$$Z = \sqrt{1700\,\Omega}$$
$$Z = 41.2\,\Omega$$

Q15. What term is given to total opposition to ac in a circuit?

Q16. What formula is used to calculate the amount of this opposition in a series circuit?

Q17. What is the value of Z in a series ac circuit where X_L = 6 ohms, X_C = 3 ohms, and R = 4 ohms?

OHMS LAW FOR AC

In general, Ohm's law cannot be applied to alternating-current circuits since it does not consider the reactance which is always present in such circuits. However, by a modification of Ohm's law which does take into consideration the effect of reactance we obtain a general law which is applicable to ac circuits. Because the impedance, Z, represents the combined opposition of all the reactances and resistances, this general law for ac is,

$$I = \frac{E}{Z}$$

this general modification applies to alternating current flowing in any circuit, and any one of the values may be found from the equation if the others are known.

For example, suppose a series circuit contains an inductor having 5 ohms resistance and 25 ohms inductive reactance in series with a capacitor having 15 ohms capacitive reactance. If the voltage is 50 volts, what is the current? This circuit can be drawn as shown in figure 4-8.

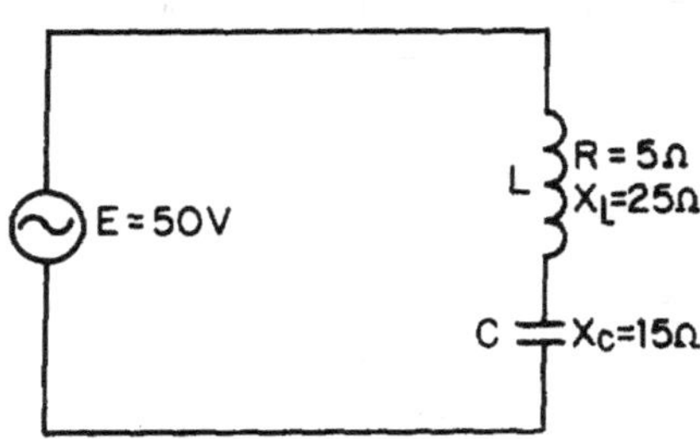

Figure 4-8.—Series LC circuit.

Given: $R = 5\,\Omega$

$X_L = 25\,\Omega$

$X_C = 15\,\Omega$

$E = 50\,V$

Solution: $X = X_L - X_C$

$X = 25\,\Omega - 15\,\Omega$

$X = 10\,\Omega$

$Z = \sqrt{R^2 + X^2}$

$Z = \sqrt{(5\,\Omega)^2 + (10\,\Omega)^2}$

$Z = \sqrt{25 + 100\,\Omega}$

$Z = \sqrt{125\,\Omega}$

$Z = 11.2\,\Omega$

$I = \frac{E}{Z}$

$I = \frac{50\,V}{11.2\,\Omega}$

$I = 4.46\,A$

Now suppose the circuit contains an inductor having 5 ohms resistance and 15 ohms inductive reactance in series with a capacitor having 10 ohms capacitive reactance. If the current is 5 amperes, what is the voltage?

Given: $R = 5\,\Omega$

$X_L = 15\,\Omega$

$X_C = 10\,\Omega$

$I = 5A$

Solution:

$$X = X_L - X_C$$
$$X = 15\,\Omega - 10\,\Omega$$
$$X = 5\,\Omega$$
$$Z = \sqrt{R^2 + X^2}$$
$$Z = \sqrt{(5\,\Omega)^2 + (5\,\Omega)^2}$$
$$Z = \sqrt{25 + 25\,\Omega}$$
$$Z = \sqrt{50\,\Omega}$$
$$Z = 7.07\,\Omega$$
$$E = IZ$$
$$E = 5A \times 7.07\,\Omega$$
$$E = 35.35V$$

Q18. What are the Ohm's law formulas used in an ac circuit to determine voltage and current?

POWER IN AC CIRCUITS

You know that in a direct current circuit the power is equal to the voltage times the current, or $P = E \times I$. If a voltage of 100 volts applied to a circuit produces a current of 10 amperes, the power is 1000 watts. This is also true in an ac circuit when the current and voltage are in phase; that is, when the circuit is effectively resistive. But, if the ac circuit contains reactance, the current will lead or lag the voltage by a certain amount (the phase angle). When the current is out of phase with the voltage, the power indicated by the product of the applied voltage and the total current gives only what is known as the APPARENT POWER. The TRUE POWER depends upon the phase angle between the current and voltage. The symbol for phase angle is θ (Theta).

When an alternating voltage is impressed across a capacitor, power is taken from the source and stored in the capacitor as the voltage increases from zero to its maximum value. Then, as the impressed voltage decreases from its maximum value to zero, the capacitor discharges and returns the power to the source. Likewise, as the current through an inductor increases from its zero value to its maximum value the field around the inductor builds up to a maximum, and when the current decreases from maximum to zero the field collapses and returns the power to the source. You can see therefore that no power is used up in either case, since the power alternately flows to and from the source. This power that is returned to the source by the reactive components in the circuit is called REACTIVE POWER.

In a purely resistive circuit all of the power is consumed and none is returned to the source; in a purely reactive circuit no power is consumed and all of the power is returned to the source. It follows that in a circuit which contains both resistance and reactance there must be some power dissipated in the resistance as well as some returned to the source by the reactance. In figure 4-9 you can see the relationship between the voltage, the current, and the power in such a circuit. The part of the power curve which is shown below the horizontal reference line is the result of multiplying a positive instantaneous value of current by a negative instantaneous value of the voltage, or vice versa. As you know, the product obtained by multiplying a positive value by a negative value will be negative. Therefore the power at that instant must be considered as negative power. In other words, during this time the reactance was returning power to the source.

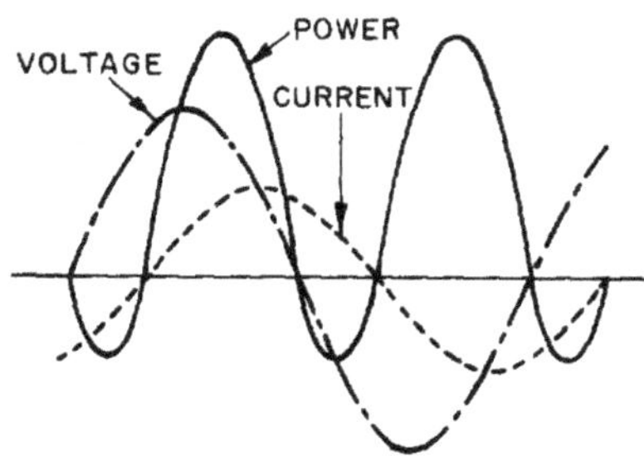

Figure 4-9.—Instantaneous power when current and voltage are out of phase.

The instantaneous power in the circuit is equal to the product of the applied voltage and current through the circuit. When the voltage and current are of the same polarity they are acting together and taking power from the source. When the polarities are unlike they are acting in opposition and power is being returned to the source. Briefly then, in an ac circuit which contains reactance as well as resistance, the apparent power is reduced by the power returned to the source, so that in such a circuit the net power, or true power, is always less than the apparent power.

Calculating True Power in AC Circuits

As mentioned before, the true power of a circuit is the power actually used in the circuit. This power, measured in watts, is the power associated with the total resistance in the circuit. To calculate true power, the voltage and current associated with the resistance must be used. Since the voltage drop across the resistance is equal to the resistance multiplied by the current through the resistance, true power can be calculated by the formula:

$$\text{True Power} = (I_R)^2 R$$

Where: True Power is measured in watts

I_R is resistive current in amperes

R is resistance in ohms

For example, find the true power of the circuit shown in figure 4-10.

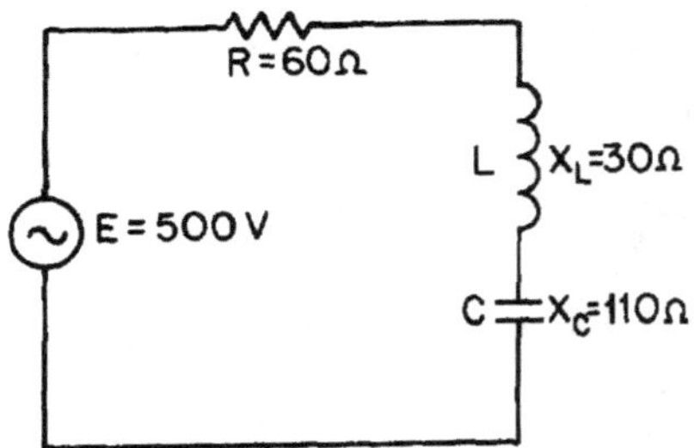

Figure 4-10.—Example circuit for determining power.

Given:

$$R = 60\ \Omega$$
$$X_L = 30\ \Omega$$
$$X_C = 110\ \Omega$$
$$E = 500\,V$$

Solution:

$$X = X_C - X_L$$
$$X = 110\Omega - 30\ \Omega$$
$$X = 80\ \Omega$$
$$Z = \sqrt{R^2 + X^2}$$
$$Z = \sqrt{(60\ \Omega)^2 + (80\ \Omega)^2}$$
$$Z = \sqrt{3600 + 6400\ \Omega}$$
$$Z = \sqrt{10{,}000\ \Omega}$$
$$Z = 100\ \Omega$$
$$I = \frac{E}{Z}$$
$$I = \frac{500\,V}{100\Omega}$$
$$I = 5A$$

Since the current in a series circuit is the same in all parts of the circuit:

$$\text{TruePower} = (I_R)^2 R$$
$$\text{TruePower} = (5\,A)^2 \times 60\ \Omega$$
$$\text{True Power} = 1500\ \text{watts}$$

Q19. What is the true power in an ac circuit?

Q20. What is the unit of measurement of true power?

Q21. What is the formula for calculating true power?

Calculating Reactive Power in AC Circuits

The reactive power is the power returned to the source by the reactive components of the circuit. This type of power is measured in Volt-Amperes-Reactive, abbreviated var.

Reactive power is calculated by using the voltage and current associated with the circuit reactance.

Since the voltage of the reactance is equal to the reactance multiplied by the reactive current, reactive power can be calculated by the formula:

$$\text{Reactive Power} = (I_X)^2 X$$

Where: Reactive power is measured in volt-amperes-reactive.

I_X is reactive current in amperes.

X is total reactance in ohms.

Another way to calculate reactive power is to calculate the inductive power and capacitive power and subtract the smaller from the larger.

$$\text{Reactive Power} = (I_L)^2 X_L - (I_C)^2 X_C$$

or

$$(I_C)^2 X_C - (I_L)^2 X_L$$

Where: Reactive power is measured in volt-amperes-reactive.

I_C is capacitive current in amperes.

X_C is capacitive reactance in ohms.

I_L is inductive current in amperes.

X_L is inductive reactance in ohms.

Either one of these formulas will work. The formula you use depends upon the values you are given in a circuit.

For example, find the reactive power of the circuit shown in figure 4-10.

Given:

$$X_L = 30\ \Omega$$
$$X_C = 110\ \Omega$$
$$X = 80\ \Omega$$
$$I = 5\,A$$

Since this is a series circuit, current (I) is the same in all parts of the circuit.

Solution:

$$\text{Reactive power} = (I_X)^2 X$$
$$\text{Reactive power} = (5\,A)^2 \times 80\ \Omega$$
$$\text{Reactive power} = 2{,}000\ \text{var}$$

To prove the second formula also works,

$$\text{Reactive power} = (I_C)^2 X_C - (I_L)^2 X_L$$
$$\text{Reactive power} = (5\,A)^2 \times 110\ \Omega - (5\,A)^2 \times 30\ \Omega$$
$$\text{Reactive power} = 2{,}750\ \text{var} - 750\ \text{var}$$
$$\text{Reactive power} = 2000\ \text{var}$$

Q22. What is the reactive power in an ac circuit?

Q23. What is the unit of measurement for reactive power?

Q24. What is the formula for computing reactive power?

Calculating Apparent Power in AC Circuits.

Apparent power is the power that appears to the source because of the circuit impedance. Since the impedance is the total opposition to ac, the apparent power is that power the voltage source "sees." Apparent power is the combination of true power and reactive power. Apparent power is not found by simply adding true power and reactive power just as impedance is not found by adding resistance and reactance.

To calculate apparent power, you may use either of the following formulas:

$$\text{Apparent power} = (I_Z)^2 Z$$

Where: Apparent power is measured in VA (volt-amperes)

I_Z is impedance current in amperes.

Z is impedance in ohms.

or

$$\text{Apparent power} = \sqrt{(\text{True power})^2 + (\text{reactive power})^2}$$

For example, find the apparent power for the circuit shown in figure 4-10.

Given: $Z = 100\ \Omega$

$I = 5\ \text{A}$

Recall that current in a series circuit is the same in all parts of the circuit.

Solution:

$$\text{Apparent Power} = (I_Z)^2 Z$$

$$\text{Apparent power} = (5\ \text{A})^2 \times 100\ \Omega$$

$$\text{Apparent power} = 2500\ \text{VA}$$

or

Given:

$\text{True power} = 1500\ \text{W}$

$\text{Reactive power} = 2000\ \text{var}$

$$\text{Apparent power} = \sqrt{(\text{True power})^2 + (\text{reactive power})^2}$$

$$\text{Apparent power} = \sqrt{(1500\ \text{W})^2 + (2000\ \text{var})^2}$$

$$\text{Apparent power} = \sqrt{625 \times 10^4}\ \text{VA}$$

$$\text{Apparent power} = 2500\ \text{VA}$$

Q25. What is apparent power?

Q26. What is the unit of measurement for apparent power?

Q27. What is the formula for apparent power?

Power Factor

The POWER FACTOR is a number (represented as a decimal or a percentage) that represents the portion of the apparent power dissipated in a circuit.

If you are familiar with trigonometry, the easiest way to find the power factor is to find the cosine of the phase angle (θ). The cosine of the phase angle is equal to the power factor.

You do not need to use trigonometry to find the power factor. Since the power dissipated in a circuit is true power, then:

$$\text{Apparent Power} \times \text{PF} = \text{True Power},$$

$$\text{Therefore,} \qquad \text{PF} = \frac{\text{True Power}}{\text{Apparent Power}}$$

If true power and apparent power are known you can use the formula shown above.

Going one step further, another formula for power factor can be developed. By substituting the equations for true power and apparent power in the formula for power factor, you get:

$$\text{PF} = \frac{(I_R)^2 R}{(I_Z)^2 Z}$$

Since current in a series circuit is the same in all parts of the circuit, I_R equals I_Z. Therefore, in a series circuit,

$$\text{PF} = \frac{R}{Z}$$

For example, to compute the power factor for the series circuit shown in figure 4-10, any of the above methods may be used.

Given:

$$\text{True Power} = 1500\text{V}$$
$$\text{Apparent Power} = 2500 \text{ VA}$$

Solution:

$$\text{PF} = \frac{\text{True Power}}{\text{Apparent Power}}$$
$$\text{PF} = \frac{1500 \text{ W}}{2500 \text{ VA}}$$
$$\text{PF} = .6$$

Another method:

Given: $R = 60\ \Omega$

$Z = 100\ \Omega$

Solution:

$$PF = \frac{R}{Z}$$

$$PF = \frac{60\ \Omega}{100\ \Omega}$$

$$PF = .6$$

If you are familiar with trigonometry you can use it to solve for angle θ and the power factor by referring to the tables in appendices V and VI.

Given: $R = 60\ \Omega$

$X = 80\ \Omega$

Solution:

$$\tan\ \theta = \frac{X}{R}$$

$$\tan\ \theta = \frac{80\ \Omega}{60\ \Omega}$$

$$\tan\ \theta = 1.333$$

$$\theta = 53.1^\circ$$

$$PF = \cos\ \theta$$

$$PF = .6$$

NOTE: As stated earlier the power factor can be expressed as a decimal or percentage. In this example the decimal number .6 could also be expressed as 60%.

Q28. What is the power factor of a circuit?

Q29. What is a general formula used to calculate the power factor of a circuit?

Power Factor Correction

The apparent power in an ac circuit has been described as the power the source "sees". As far as the source is concerned the apparent power is the power that must be provided to the circuit. You also know that the true power is the power actually used in the circuit. The difference between apparent power and true power is wasted because, in reality, only true power is consumed. The ideal situation would be for apparent power and true power to be equal. If this were the case the power factor would be 1 (unity) or 100 percent. There are two ways in which this condition can exist. (1) If the circuit is purely resistive or (2) if the circuit "appears" purely resistive to the source. To make the circuit appear purely resistive there must be no reactance. To have no reactance in the circuit, the inductive reactance (X_L) and capacitive reactance (X_C) must be equal.

Remember: $X = X_L - X_c$

Therefore, when

$$X_L = X_C,\ X = 0$$

The expression "correcting the power factor" refers to reducing the reactance in a circuit.

The ideal situation is to have no reactance in the circuit. This is accomplished by adding capacitive reactance to a circuit which is inductive and inductive reactance to a circuit which is capacitive. For example, the circuit shown in figure 4-10 has a total reactance of 80 ohms capacitive and the power factor was .6 or 60 percent. If 80 ohms of inductive reactance were added to this circuit (by adding another inductor) the circuit would have a total reactance of zero ohms and a power factor of 1 or 100 percent. The apparent and true power of this circuit would then be equal.

Q30. An ac circuit has a total reactance of 10 ohms inductive and a total resistance of 20 ohms. The power factor is .89. What would be necessary to correct the power factor to unity?

SERIES RLC CIRCUITS

The principles and formulas that have been presented in this chapter are used in all ac circuits. The examples given have been series circuits.

This section of the chapter will not present any new material, but will be an example of using all the principles presented so far. You should follow each example problem step by step to see how each formula used depends upon the information determined in earlier steps. When an example calls for solving for square root, you can practice using the square-root table by looking up the values given.

The example series RLC circuit shown in figure 4-11 will be used to solve for X_L, X_C, X, Z, I_T, true power, reactive power, apparent power, and power factor.

The values solved for will be rounded off to the nearest whole number.

First solve for X_L and X_C.

Given: $f = 60\,Hz$
$L = 27\,mH$
$C = 380\,\mu F$

Solution:
$$X_L = 2\pi fl$$
$$X_L = 6.28 \times 60\,Hz \times 27\,mH$$
$$X_L = 10\ \Omega$$
$$X_C = \frac{1}{2\pi fc}$$
$$X_C = \frac{1}{6.28 \times 60\,Hz \times 380\,\mu F}$$
$$X_C = \frac{1}{0.143}\Omega$$
$$X_C = 7\,\Omega$$

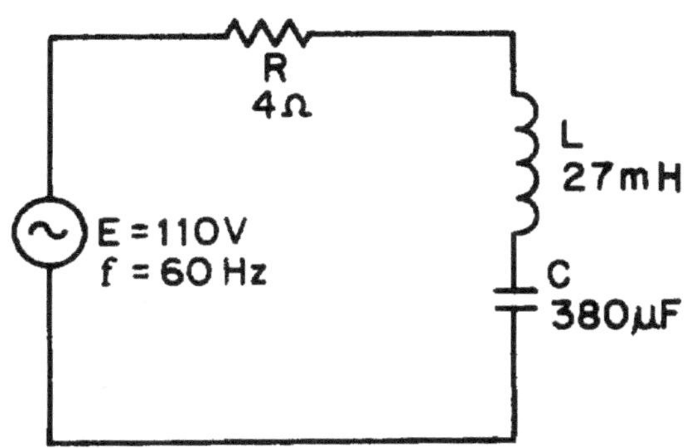

Figure 4-11.—Example series RLC circuit

Now solve for X

Given: $X_C = 7\ \Omega$
$X_L = 10\ \Omega$

Solution:
$$X = X_L - X_C$$
$$X = 10\,\Omega - 7\,\Omega$$
$$X = 3\,\Omega \text{ (Inductive)}$$

Use the value of X to solve for Z.

Given: $X = 3\ \Omega$

$R = 4\ \Omega$

Solution:

$$Z = \sqrt{X^2 + R^2}$$

$$Z = \sqrt{(3\ \Omega)^2 + (4\ \Omega)^2}$$

$$Z = \sqrt{9 + 16\ \Omega}$$

$$Z = \sqrt{25\ \Omega}$$

$$Z = 5\Omega$$

This value of Z can be used to solve for total current (I_T).

Given: $Z = 5\Omega$

$E = 110\ V$

Solution:

$$I_T = \frac{E}{Z}$$

$$I_T = \frac{110\ V}{5\Omega}$$

$$I_T = 22\ A$$

Since current is equal in all parts of a series circuit, the value of I_T can be used to solve for the various values of power.

Given: $I_T = 22\text{ A}$

$R = 4\Omega$

$X = 3\Omega$

$Z = 5\Omega$

Solution:

$$\text{True Power} = (I_R)^2 R$$
$$\text{True Power} = (22\text{A})^2 \times 4\Omega$$
$$\text{True Power} = 1936\text{ W}$$

$$\text{Reactive power} = (I_X)^2 X$$
$$\text{Reactive power} = (22\text{A})^2 \times 3\Omega$$
$$\text{Reactive power} = 1452\text{ var}$$

$$\text{Apparent power} = (I_Z)^2 Z$$
$$\text{Apparent Power} = (22\text{A})^2 \times 5\Omega$$
$$\text{Apparent Power} = 2420\text{ VA}$$

The power factor can now be found using either apparent power and true power or resistance and impedance. The mathematics in this example is easier if you use impedance and resistance.

Given: $R = 4\Omega$

$Z = 5\Omega$

Solution:

$$PF = \frac{R}{Z}$$
$$PF = \frac{4\Omega}{5\Omega}$$
$$PF = .8 \text{ or } 80\%$$

PARALLEL RLC CIRCUITS

When dealing with a parallel ac circuit, you will find that the concepts presented in this chapter for series ac circuits still apply. There is one major difference between a series circuit and a parallel circuit that must be considered. The difference is that current is the same in all parts of a series circuit, whereas voltage is the same across all branches of a parallel circuit. Because of this difference, the total impedance of a parallel circuit must be computed on the basis of the current in the circuit.

You should remember that in the series RLC circuit the following three formulas were used to find reactance, impedance, and power factor:

$$X = X_L - X_C \text{ or } X = X_C - X_L$$
$$Z = \sqrt{(I_R)^2 + X^2}$$
$$PF = \frac{R}{Z}$$

When working with a parallel circuit you must use the following formulas instead:

$$I_X = I_L - I_C \text{ or } I_X = I_C - I_L$$
$$I_Z = \sqrt{(I_R)^2 + (I_X)^2}$$
$$PF = \frac{I_R}{I_Z}$$

(The impedance of a parallel circuit is found by the formula $Z = \frac{E}{I_Z}$)

NOTE: If no value for E is given in a circuit, any value of E can be assumed to find the values of I_L, I_C, I_X, I_R, and I_Z. The same value of voltage is then used to find impedance.

For example, find the value of Z in the circuit shown in figure 4-12.

Given:

$$E = 300 \text{ V}$$
$$R = 100\ \Omega$$
$$X_L = 50\ \Omega$$
$$X_C = 150\ \Omega$$

The first step in solving for Z is to calculate the individual branch currents.

Solution:

$$I_R = \frac{E}{R}$$

$$I_R = \frac{300\ V}{100\ \Omega}$$

$$I_R = 3\ A$$

$$I_L = \frac{E}{X_L}$$

$$I_L = \frac{300\ V}{50\ \Omega}$$

$$I_L = 6\ A$$

$$I_C = \frac{E}{X_C}$$

$$I_C = \frac{300\ V}{150\ \Omega}$$

$$I_C = 2\ A$$

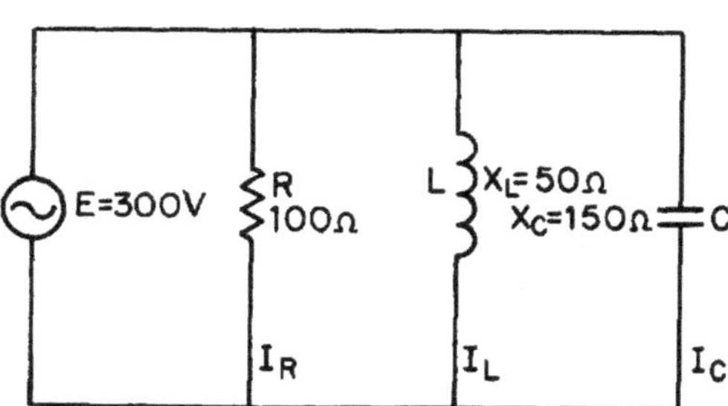

Figure 4-12.—Parallel RLC circuit.

Using the values for I_R, I_L, and I_C, solve for I_X and I_Z.

$$I_X = I_L - I_C$$

$$I_X = 6\ A - 2A$$

$$I_X = 4\ A\ (\text{inductive})$$

$$I_Z = \sqrt{(I_R)^2 + (I_X)^2}$$

$$I_Z = \sqrt{(3\ A)^2 + (4A)^2}$$

$$I_Z = \sqrt{25\ A}$$

$$I_Z = 5\ A$$

Using this value of I_Z, solve for Z.

$$Z = \frac{E}{I_Z}$$

$$Z = \frac{300\ V}{5\ A}$$

$$Z = 60\ \Omega$$

If the value for E were not given and you were asked to solve for Z, any value of E could be assumed. If, in the example problem above, you assume a value of 50 volts for E, the solution would be:

Given:

$$R = 100\ \Omega$$

$$X_L = 50\ \Omega$$

$$X_C = 150\ \Omega$$

$$E = 50\ V\ \text{(assumed)}$$

First solve for the values of current in the same manner as before.

Solution:

$$I_R = \frac{E}{R}$$

$$I_R = \frac{50\ V}{100\ \Omega}$$

$$I_R = .5\ A$$

$$I_L = \frac{E}{X_L}$$

$$I_L = \frac{50\ V}{50\ \Omega}$$

$$I_L = 1\ A$$

$$I_C = \frac{E}{X_C}$$

$$I_C = \frac{50\ V}{150\ \Omega}$$

$$I_C = .33\ A$$

Solve for I_X and I_Z.

$$I_X = I_L - I_C$$
$$I_X = 1A - .33A$$
$$I_X = .67A \text{ (Inductive)}$$
$$I_Z = \sqrt{(I_R)^2 + (I_X)^2}$$
$$I_Z = \sqrt{(0.5A)^2 + (0.67A)^2}$$
$$I_Z = \sqrt{0.6989A}$$
$$I_Z = 0.836A$$

Solve for Z.

$$Z = \frac{E}{I_Z}$$
$$Z = \frac{50\text{ V}}{.836A}$$
$$Z = 60\ \Omega \text{ (rounded off)}$$

When the voltage is given, you can use the values of currents, I_R, I_X, and I_Z, to calculate for the true power, reactive power, apparent power, and power factor. For the circuit shown in figure 4-12, the calculations would be as follows.

To find true power,

Given: $R = 100\ \Omega$
$I_R = 3A$

Solution:

$$\text{True Power} = (I_R)^2 X$$
$$\text{True Power} = (3A)^2 \times 75\ \Omega$$
$$\text{True Power} = 900\text{ W}$$

To find reactive power, first find the value of reactance (X).

Given: $E = 300\ V$

$I_X = 4\ A$ (Inductive)

Solution:

$$X = \frac{E}{I_X}$$

$$X = \frac{300\ V}{4A}$$

$$X = 75\ \Omega \text{ (Inductive)}$$

$$\text{Reactive power} = (I_X)^2 X$$

$$\text{Reactive power} = (4A)^2 \times 75\ \Omega$$

$$\text{Reactive power} = 1200\ \text{var}$$

To find apparent power,

Given: $Z = 60\ \Omega$

$I_Z = 5A$

Solution:

$$\text{Apparent Power} = (I_Z)^2 Z$$

$$\text{Apparent Power} = (5A)^2 \times 60\ \Omega$$

$$\text{Apparent Power} = 1500\ \text{VA}$$

The power factor in a parallel circuit is found by either of the following methods.

Given:

$$\text{True Power} = 900\ \text{W}$$
$$\text{Apparent Power} = 1500\ \text{VA}$$

Solution:

$$PF = \frac{\text{true power}}{\text{apparent power}}$$
$$PF = \frac{900\ \text{W}}{1500\ \text{VA}}$$
$$PF = .6$$

or

Given:

$$I_R = 3\ \text{A}$$
$$I_Z = 5\ \text{A}$$

Solution:

$$PF = \frac{I_R}{I_Z}$$
$$PF = \frac{3\ \text{A}}{5\ \text{A}}$$
$$PF = .6$$

Q31. What is the difference between calculating impedance in a series ac circuit and in a parallel ac circuit?

SUMMARY

With the completion of this chapter you now have all the building blocks for electrical circuits. The subjects covered from this point on will be based upon the concepts and relationships that you have learned. The following summary is a brief review of the subjects covered in this chapter.

INDUCTANCE IN AC CIRCUITS—An inductor in an ac circuit opposes any change in current flow just as it does in a dc circuit.

PHASE RELATIONSHIPS OF AN INDUCTOR—The current lags the voltage by 90° in an inductor (ELI).

INDUCTIVE REACTANCE—The opposition an inductor offers to ac is called inductive reactance. It will increase if there is an increase in frequency or an increase in inductance. The symbol is X_L, and the formula is $X_L = 2\pi fL$.

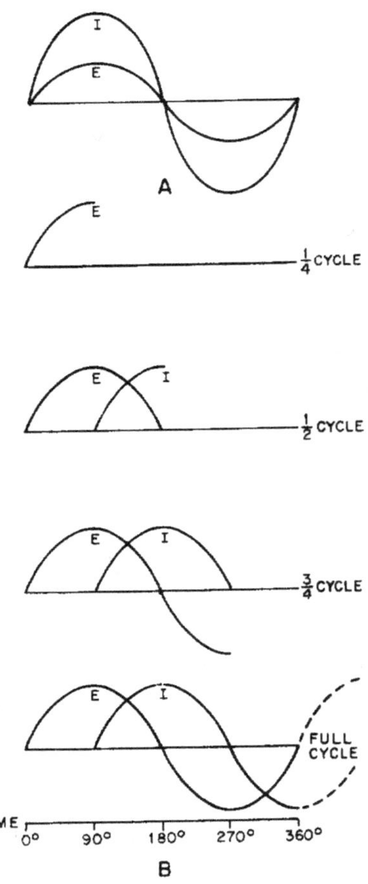

CAPACITANCE IN AC CIRCUITS—A capacitor in an ac circuit opposes any change in voltage just as it does in a dc circuit.

PHASE RELATIONSHIPS OF A CAPACITOR—The current leads the voltage by 90° in a capacitor (ICE).

CAPACITIVE REACTANCE—The opposition a capacitor offers to ac is called capacitive reactance. Capacitive reactance will decrease if there is an increase in frequency or an increase in capacitance. The symbol is X_C and the formula is

$$X_C = \frac{1}{2\pi f C}$$

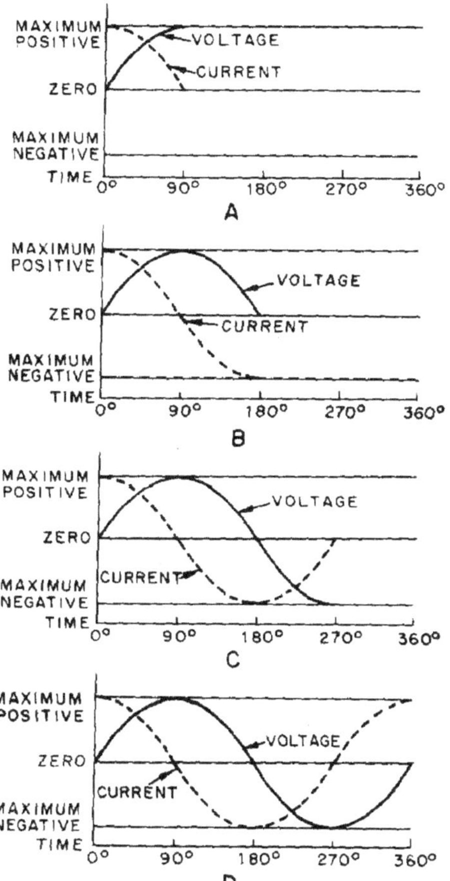

TOTAL REACTANCE—The total reactance of a series ac circuit is determined by the formula $X = X_L - X_C$ or $X = X_C - X_L$. The total reactance in a series circuit is either capacitive or inductive depending upon the largest value of X_C and X_L. In a parallel circuit the reactance is determined by

$$\frac{E}{I_X},$$

where $I_X = I_C - I_L$ or $I_X = I_L - I_C$. The reactance in a parallel circuit is either capacitive or inductive depending upon the largest value of I_L and I_C.

IMPEDANCE – The total opposition to a.c. is called impedance. The symbol is Z. In a series circuit $Z = \sqrt{R^2 + X^2}$. In a parallel circuit $I_Z = \sqrt{(I_R)^2 + (I_X)^2}$ and $Z = \frac{E}{I_Z}$.

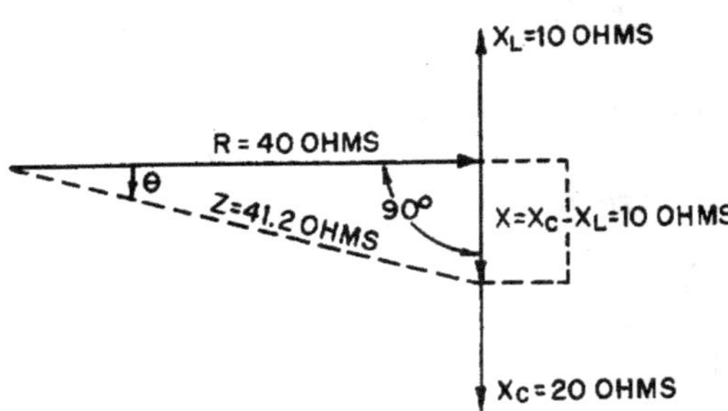

PHASE ANGLE—The number of degrees that current leads or lags voltage in an ac circuit is called the phase angle. The symbol is θ.

OHM'S LAW FORMULAS FOR AC—The formulas derived for Ohm's law used in ac are: $E = IZ$ and $I = E/Z$.

TRUE POWER—The power dissipated across the resistance in an ac circuit is called true power. It is measured in watts and the formula is: True Power = $(I_R)^2R$.

REACTIVE POWER—The power returned to the source by the reactive elements of the circuit is called reactive power. It is measured in volt-amperes reactive (var). The formula is: Reactive Power = $(I_X)^2X$.

APPARENT POWER—The power that appears to the source because of circuit impedance is called apparent power. It is the combination of true power and reactive power and is measured in volt-amperes (VA). The formulas are:

$$\text{Apparent Power} = (I_Z)^2 Z$$

$$\text{Apparent Power} = \sqrt{(\text{true power})^2 + (\text{reactive power})^2}$$

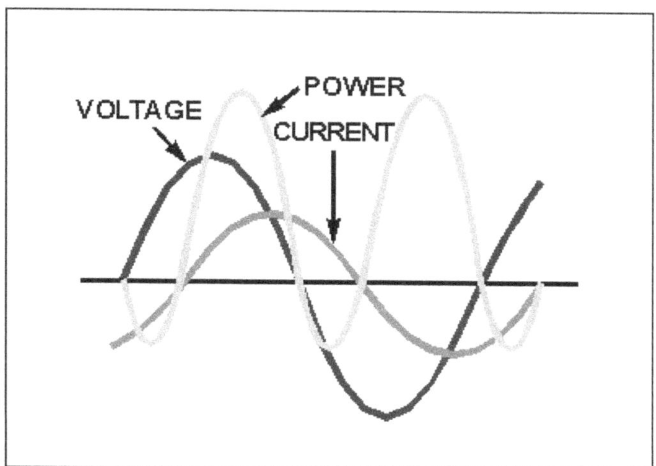

POWER FACTOR—The portion of the apparent power dissipated in a circuit is called the power factor of the circuit. It can be expressed as a decimal or a percentage. The formulas for power

factor are $PF = \frac{\text{true power}}{\text{apparent power}}$ or $PF = \cos\theta$. In a series circuit, $PF = \frac{R}{Z}$. In a parallel circuit, $Pf = \frac{I_R}{I_Z}$.

POWER FACTOR CORRECTION—To reduce losses in a circuit the power factor should be as close to unity or 100% as possible. This is done by adding capacitive reactance to a circuit when the total reactance is inductive. If the total reactance is capacitive, inductive reactance is added in the circuit.

ANSWERS TO QUESTIONS Q1. THROUGH Q31.

A1. An inductor opposes a change in current.

A2. Current lags voltage by 90° (ELI).

A3. Inductive reactance.

A4. $X_L = 2\pi fL$.

A5. X_L *increases.*

A6. X_L *decreases.*

A7. The capacitor opposes any change in voltage.

A8. Current leads voltage by 90° (ICE).

A9. Capacitive reactance.

A10.

$$X = \frac{1}{2\pi fC}$$

A11. X_C *increases.*

A12. X_C *decreases.*

A13. $X = X_L - X_C$ *or* $X = X_C - X_L$

A14. 30 Ω (capacitive).

A15. Impedance.

A16.

$$Z = \sqrt{R^2 + X^2}.$$

A17. $Z = 5\Omega$.

A18.

$$E = IZ$$

$$I = \frac{E}{Z}.$$

A19. True power is the power dissipated in the resistance of the circuit or the power actually used in the circuit.

A20. Watt.

A21. True Power = $(I_R)^2R$.

A22. Reactive power is the power returned to the source by the reactive components of the circuit.

A23. var.

A24.

$$\text{Reactive Power} = (I_X)^2 X \text{ or}$$

$$(I_C)^2 X_C - (I_L)^2 X_L \text{ or}$$

$$(I_L)^2 X_L - (I_C)^2 X_C.$$

A25. The power that appears to the source because of circuit impedance, or the combination of true power and reactive power.

A26. VA (volt-amperes).

A27.

$$\text{Apparent power} = (I_Z)^2 Z \text{ or}$$

$$\sqrt{(\text{true power})^2 + (\text{reactive power})^2}$$

A28. PF is a number representing the portion of apparent power actually dissipated in a circuit.

A29.

$$PF = \frac{\text{true power}}{\text{apparent power}} \text{ or } PF = \cos\theta.$$

A30. Add 10 ohms of capacitive reactance to the circuit.

A31. In a series circuit impedance is calculated from the values of resistance and reactance. In a parallel circuit, the values of resistive current and reactive current must be used to calculate total current (impedance current) and this value must be divided into the source voltage to calculate the impedance.

NOTES

NOTES

NOTES

NOTES

NOTES

NOTES

NOTES

NOTES

www.ingramcontent.com/pod-product-compliance
Lightning Source LLC
Chambersburg PA
CBHW081112290526
45795CB00006B/2100

* 9 7 9 8 8 7 8 0 5 0 8 3 8 *